MW01421597

PATHS TO NOWHERE

Africa's Endless walk to Economic Freedom

PATHS TO NOWHERE

Africa's Endless walk to Economic Freedom

SAMUEL APPIAH-KUBI

Copyright © 2019 by Samuel Appiah-Kubi.

	Library of Congress Control Number:	2019902545
ISBN:	Hardcover	978-1-7960-1898-1
	Softcover	978-1-7960-1897-4
	eBook	978-1-7960-1896-7

All rights reserved. No part of this book may be reproduced or transmitted in any form or by any means, electronic or mechanical, including photocopying, recording, or by any information storage and retrieval system, without permission in writing from the copyright owner.

The views expressed in this work are solely those of the author and do not necessarily reflect the views of the publisher, and the publisher hereby disclaims any responsibility for them.

Any people depicted in stock imagery provided by Getty Images are models, and such images are being used for illustrative purposes only.
Certain stock imagery © Getty Images.

Print information available on the last page.

Rev. date: 03/14/2019

To order additional copies of this book, contact:
Xlibris
1-888-795-4274
www.Xlibris.com
Orders@Xlibris.com
791545

Contents

Preface vii
Chapter 1 Hope: The Inspirational Path 1
Chapter 2 Resource Waste: The Path Of Poverty 10
Chapter 3 Governance Institutions: The Path Of Disorder 22
Chapter 4 Lawlessness: The Bumpy Path 38
Chapter 5 Dependency Mind-Set: The Path Of Stagnation 49
Chapter 6 Education: The Path To Aliens Citizenry 57
Chapter 7 Corruption: The Path Of Individualism And Division 74
Chapter 8 Technology And Data Management: The Necessary But Neglected Path 91
Chapter 9 Donor Partners: The Borrowed Path 107
Chapter 10 The African Dream: The Promising Paths 119
Referencing 129

PREFACE

As the full moon lights up the African night skies, many people in the not quite distant past will gather to listen to stories from older folks. From tales and myths to religious stories and history, the stories told in the cool of the night was what every child in the towns and villages looked forward to. These stories helped to entertain and educate the African child. A few decades ago, I used to sit among the young people in the New Atwene area who gathered at the compound of my late father to listen to such stories.

My favorite stories were the stories of colonization in Africa and the struggle for liberation from colonial governments. I was particularly enthralled by the plight of the African under colonial rule and the resistance from the Asante Kingdom that, fought and at times defeated the white man who tried to establish their form of governance and control over my tribesmen. Stories about how the black man was treated in his own land were horrifying but intriguing to hear. What was more inspiring was the bravery of those who resisted the 'oppressors' rule.

The courage and wisdom of the freedom fighters was captivating. The support these courageous men received from people in far and nearby countries where they used to run to in times of persecution were captivating. Of much interest to me were the reasons behind the independence struggle and why they chose to risk their lives and often spend years in exile. To those who had the opportunity to listen to the stories, the love and commitment demonstrated by these brave men to the land Africa and the continent's future generation were so inspirational.

The memory of the life under colonial rule, the struggle for independence and the process of independence was so fresh on my father's mind that a lot of the moonlight stories centered on late years of the Gold Coast and the early years of modern Ghana. He made us understand that he was part of the strikes, the protests, the voting, and the non-violent activities that culminated in the independence of the Gold Coast-Modern day Ghana- in 1957. He played an active part in the process because of us-the younger folks. He admits that, though he was not in the frontlines of the struggle, he was with the masses that were prepared to give up their lives so that the generations after them would have better lives.

As common to all oral stories, what happened may have been exaggerated, understated or told as it was. What remains factual is the control from foreigners on the land of the black man in the colonial period and the euphoria that greeted the independence of the nations of Africa and their resolve to build more prosperous countries. The welcoming knowledge that the gold resources of the former Gold Coast were now in the hands of Ghanaians and the expectations of a brighter future for Ghana was on my mind as I grew up. It even influenced my choice of course of study at the university as I concentrated on economics and political science because I was inspired by my father's stories of old to acquire knowledge to help in the building of our nation and to assist in the equitable distribution of our nation's resources.

Fast forward to modern times, Africa has had many post-independence success stories. However, the sentiments being expressed by most Africans especially the youth reveals that all is not well with African countries as many are economically weak and engulfed in unsustainable debts. The higher percentage of the reportedly booming economic activities of many African countries do not accrue to the people of Africa but migrants from other continents. Only a few Africans are able to tap into the economic activities of their lands, a situation that is widening the income gap and sending the more significant percentage of the population into poverty. Most economies are not expanding enough for the growing population raising unemployment to unprecedented levels. The statistics are becoming scary with each passing day and fast approaching the tipping point.

Many who are unable to tap into Africa's prosperity are resorting to all forms of activities that can best be described as desperate attempts for survival. The desperate attempts are evident in many ways and all of them are threatening to the peace and security of Africa.

Engaging in criminal activities, violent demonstrations, joining militants and terror groups and embarking on dangerous journeys to countries that seem to offer better opportunities are some of the desperate attempts.

I was motivated to write this piece because it appears the leadership in Africa do not appreciate the gravity of the situation. If they do, then there may be something wrong with the course of action that is being taken to address the multi-faceted challenge. Though the leadership may be aware of the situation, I hope this little piece might cause them to have a second look at the factors they might have overlooked or deliberately ignored. I also hope this piece will inspire the ordinary African who have already given up and are throwing their hands in despair to arise and play their part in addressing the present challenge because they have consciously or unconsciously been part of the of the problem. They remained silent while elected leaders exercised powers they have including the powers they don't have over them. They prefer 'gambling' with shortcuts to success rather than going through the whole process which though has extended gestation period, has a higher success rate.

Economic independence is within reach of many African countries, but the actions on the part of the rulers and the inactions of the ruled may perpetuate the status quo. I fear that the situation might get worse if the right actions are deferred. I believe this piece will motivate the leaders to take the much-needed actions no matter how tough it may be to address the concerns of the present generation for the good of future generations. I also believe the masses will better appreciate any seemingly uncomfortable actions that their leaders may take in the interest of their respective countries and be patient for prudent and sustainable measures to take roots instead of the usual quick, superficial programs. In a globalized world, I also believe other players outside the continent of Africa will also support Africa's efforts at building stronger economies as the economic success of African States holds immeasurable benefits to the world.

I have had the opportunity to travel to many countries in Africa and have interacted with Africans from different backgrounds and ethnic groupings. Moreover, on the airplanes, I have sat next to bound and loosed African deportees, and I have heard the returning migrants narrate their ordeals. I have listened to the African speak. I have listened to the stories of Africans who are successful in various sectors of the economy, and I have heard the stories of the disadvantaged and poor African. Though differences exist, there are many similarities. In my opinion, many of the discussion with the advantaged have greedy undertones, while many of the disadvantaged want a decent life.

I am more touched by the stories of the disadvantaged because it reminds me of the stories under the moonlights of Akomadan some years ago. My father's generation shared similar frustrations under the colonial government. Like the story of the animal farm, the circumstances that called for joint efforts to overthrow the oppressor has re-emerged, probably in its worst form under the administration of fellow freedom fighters. Paths to nowhere is a reflection of the issues that are working against the African man's hope of better life echoed from the distant voices along the pathways in rural Africa to the cries that are often lost in the noisy and busy streets of urban Africa.

To serve as a public officer and to be served by a public officer in Ghana and by extension in Africa are two different experiences and I have had the opportunity to experience both. Paths to Nowhere is written because the African situation is a ticking time bomb that is ready to explode. Unless something is urgently done to reverse the trend, an explosion is imminent, and the consequences might be disastrous. A cue from the Arab springs, suggests that, when all the remote causes are set, it takes just one action such as a frustrated person setting himself on fire to set in motion series of uncontrollable events. From all indications, the remote justification for disastrous events may be established, and a spark is all that is needed to trigger an explosion.

This book is not intended to attack any individual, group of people, country, government, institution, or system. It is meant to bring the issues on the mind of many Africans to the attention of all Africans and the global community so that together, we can support Africa to fulfill her dreams and probably to avert a potential threat to global peace and security.

CHAPTER ONE

Hope: The Inspirational Path

You have to dream before your dreams can come true.
—A. P. J. Abdul Kalam

GHANA: AN EXEMPLARY COUNTRY ON A NEW PATH

After years of colonial rule in Africa, Ghana became the first black African country to attain independence from the British colonial rule in 1957. Hopes were high as well as prospects. The country had abundant natural resources, including timber, gold, bauxite, diamonds, manganese, limestone, and several others. The postindependence euphoria was high as Ghana carried the hope of black Africa. Indeed, at independence, the first president of Ghana, Dr. Kwame Nkrumah, in delivering the first independence speech at the independence square in Accra on March 6, 1957, made bold declarations that attracted the attention of Africa's friends and enemies. He affirmed that Ghana was going to be a proof to the rest of the world that the black man can independently manage a country for the good of the citizenry.

At the center of Ghana's new flag, a black star is boldly embossed. To those who laid the foundation stones of the new black African state, the black star was an indication that the country was to be the shining star of Africa, an exemplary nation that the rest of black Africa will from thence look up to and be inspired by its political and economic freedom so they can also fight for their independence. President Nkrumah advocated that Ghana's freedom "is meaningless unless it is linked up with the total liberation of Africa." Earlier, on the eve of Ghana's independence, he had declared that Ghana's independence was to prove to the world that "the black man is capable of managing his own affairs."[A6]

What it meant was that Ghana, which had attained political independence, was now treading on a new path—a path toward a different kind of independence, which was to attain economic independence while inspiring and supporting her other African counterparts to fight for their version of political independence. The inspiration yielded fruitful results as more than thirty other African countries followed the steps of Ghana and fought their way to political independence within the next decade.

Ghana's ability to overcome the struggle and free herself from colonial rule and form its government and systems was celebrated by Africans both on the continent and in the Diaspora. The Africans in the Diaspora, who were also fighting their kind of oppression outside the continent of Africa, joined in the celebration and found it necessary to take a pilgrimage to the land that has given so much hope to Africans and people of African descent. Martin Luther King Jr. and Malcolm X, who at that time were playing leading roles in the fight for equal rights for black America, found it necessary to visit the new black African state. W. E. B. DuBois, an American civil rights activist and a Pan-African, immigrated to Ghana and chose to make Ghana his new home, even after death.

The hope and belief of the people in the new black African state were kindled mainly because of the notion that their taxes were no longer going to serve the interest of the oppressive white leaders but was going to be used to enhance the welfare of the black people. The revenues from the country's rich mineral fields were going to be used to develop the infrastructure that will provide the welfare needs of the Ghanaian. The cocoa and other farm products from the fertile soils of the country were to be harnessed to the benefit of the African.

The whole setup was that Ghana was on the path toward realizing the African dream. When Kwame Nkrumah took over from the colonial rule, he embarked on massive industrialization in an attempt to reduce the country's reliance on imports and also to produce goods that Ghana was too poor at that time to import. Well-designed economic development plans were pursued and investments made. Industries were set up to process what the country produces and other industries set up to provide raw materials to feed other sectors.

The Volta River Dam was to provide not only the energy needs of the country but also the energy needs of other neighboring countries. The Tema Harbour, Tema Metropolis, and Tema Motorway were projects that opened the country up for international trade. The Adomi Bridge was constructed to integrate the Volta Region to other parts of the country. Ghana was on the right path to her economic independence to the admiration of the whole world. So impressive were Ghana's strides and smooth was the path that the World Bank in 1960 projected that in just two decades, Ghana would be a first-class economy.

A DREAM BECOMES A NIGHTMARE

Fast forwarding to today, where does Ghana stand in the comity of nations? Did Ghana become the first world economy in the 1980s as predicted by the World Bank? Was Ghana able to show to the way and lead other African nations along the path of prosperity? Did Ghana provide the needed inspiration to other African countries?

On March 26, 2018, the *Business and Financial Times* newspaper in Ghana published a disturbing report of a survey that was carried out in Ghana and other countries in sub-Saharan Africa. The research that was carried out by the Pew Research Center (PRC), a U.S.-based think tank, revealed that a staggering 75 percent of Ghanaians "would not hesitate to grab an opportunity to migrate abroad to better their living conditions."[A1] Similarly, percentages were recorded in Kenya, Nigeria, and several other African countries.

Key factors fueling the eagerness to leave the country, according to the publication, were unemployment, lower wages, and the admission on the part of the African respondents that the job market is likely to get worse in the near future. In 2015, 1.7 million Ghanaians applied to leave the country through the United States of America's Diversity Lottery Program against the total number of about fifty-five thousand the United States is seeking from the entire world for the whole year. In 2017, a total of 1.9 million Ghanaians applied for the U.S. Diversity Lottery Visa Program.[A2]

The report further revealed that Ghanaians and the rest of the sub-Saharan African population would leave their country at any possible opportunity with the majority choosing to do that by any means necessary even if it is illegal, unsafe, and life-threatening. Many wants to ride the rough waters of the Mediterranean to seek asylum in the European Union. In the years 2010-2017 the European Statistical Agency Eurostat indicated that almost one million applications for asylum were received from sub-Saharan Africans.[A3]

The results of the survey tell something more than just the figures. It's a testimony that the black man has not been able to manage his own affairs to the satisfaction of the citizenry. The citizens of Ghana who wanted to send the white man away so they can dwell safely and peaceful on their land are now ready to leave their country and sell themselves again into slavery because slavery in foreign countries now looks more promising than freedom on their motherland. The sad cries of Ghanaians and for a greater extent the African youth are echoed by the increasing number of the youth who make the long, perilous, and arduous journey from Ghana through Burkina Faso, Mali, Niger, to Libyan shores where they wait anxiously for favorable weather on the Mediterranean Sea to make it to a more prosperous Europe.

Quite recently, six abled Ghanaians between the ages of twenty-four and forty went to hide in a shipping container and nearly suffocated to death in their attempts to stow away to an unknown destination. They didn't know where the vessel that will carry the container will be heading; they just wanted to leave the shores of the country. The actions of the young men meant that anyplace on any part of the world was better for them than Ghana. During the interrogation, they revealed that they didn't know where the container will take them, but one port official just gave them the hint that the vessel that will carry that shipping container will go to Europe, but they were not sure. They were just fortunate to have sent into the container tools that they intended to use in cutting the container in case the owner of the container chooses not to open it at its final destination. They began to suffocate after three days in the container. They finally cut open the container when they could no longer contain the heat or get enough oxygen.

Where did Africa miss it? The Koreans had independence just around the same time Ghana had her freedom, but the Koreans have a better story to tell. Did Africa miss the path to economic freedom? Was the climb too steep? Or Africa didn't have the energies or capabilities to succeed?

Indeed, in the course of my extensive travels to many countries in Africa, Asia, North America, and Europe, I have heard a lot of unpleasant statements about Africa from different groups of people from different backgrounds. But a recent remark that I read in the news about my country, Ghana, and Nigeria that was purported to have come from the North Korean leader Kim Jong-Un was an eye-opener. Though the report was later tagged as 'fake news' because no substantive evidence is available, the rate at which the news went viral on social media platforms in Africa caught my attention. The fact is that it is after reading such remarks about my continent and the controversy it raised in Ghana and Nigeria, that prompted me to start writing this book. It was an article written by Joseph Nii Ankrah at the online news site of Happy FM on November 28, 2018, and it had the title; "Give me Ghana, Nigeria to recolonize; in a year, they will become first-class countries."[A4]

According to the article, the North Korean president allegedly requested that Nigeria should voluntarily surrender for recolonization, and Ghana should also give in and allow North Korea to colonize the country for the second time so that Ghana and Nigeria will eventually learn how to run a country. In essence, whether the North Korean leader made those statements or not, the facts remain untouched. Ghana and Nigeria have all the resources to make the countries the envy of the whole world, but it can't simply do that. If we could, Africa wouldn't be in this state after six decades; if we could, our citizens will not be seeking asylum in other countries while there are no disasters or wars in our countries.

When the "shit hole" statement came out, the leaders of Africa responded with a statement from the African Union Secretariat, which needs to be commended. But it has been over a year since these words; what actions are being taken to put Africa in a position that will not warrant such group of words again? The reaction of the masses to the remarks attributed to Kim Jong-Un is an indication that Africans are tired of name-calling. But the question that remains unanswered is whether the leadership in Africa, like the masses, really care about the negative name and perception about their respective countries in the other parts of the world.

I say to the good people of Ghana and Nigeria; it is time to wake up to the sense of the many unfortunate voices coming from beyond the borders of Africa. It is about time Africa had a referendum regarding recolonization in our respective countries. I wasn't surprised when the people of Britain voted yes to Brexit, and I will be the least surprised if Ghanaians or Nigerians or other African countries will vote 'yes' for recolonization at a national referendum. The people of Israel told Moses that they wanted to go back into slavery in Egypt because life in the wilderness was worse than their life in slavery. The people of Africa, particularly the youth who constitute a greater percentage of the population, can no more bear the harsh "economic wilderness" and, by actions, are calling on the leadership to take them back to slavery.

If African leaders want to test how bad they have failed their people, they should organize a free and fair referendum on recolonization as a litmus test. I am almost certain the people will vote in favor of recolonization, and this should not be taken lightly. The time for Africa's leadership to act is now because the masses are in desperate need for solutions and nothing else. The signs of desperation are beginning to show, and if recolonization presents the answer, many will opt for it. If our leaders are of the view that recolonization is nonsense, they should prove to the contrary or at least begin to take the possibility of citizens demanding that option seriously. Let Africa do the mathematics so we can appreciate the gains and losses. Ghana was a British colony for sixty-three years. In just a few months, Ghana will be sixty-three years as an independent state. The time for comparative assessment is now!

It is high time African leaders understood that the sovereignty of their countries is becoming relevant only for the leaders. To the ordinary citizens of Africa, they are ready to sacrifice that sovereignty because it is of no use to them anymore. The Akans have an adage, "sankɔfa yɛnnkyi," which means "it is not a taboo to go back and make right the wrongs." After all, African countries will not be the only countries that have taken a direction and retreated. The British got involved with the European Union, and when things were not going as they envisaged, they are on the way back to do what they think is good for the country.

The people of Africa had the hopes; they had the beliefs, they had the motivation, they had the commitment, and they had the leadership. Over the years, Africa has lost leadership, though the people have been obedient citizens. Africa's leaders don't seem to be leading. They appear to be following than leading, and if they are indeed following, whom are they following, and haven't they realized that they are being misled along the wrong paths? Some of Africa's leaders have used words as sedatives to keep their people calm and patient; they have always preached that things will be all right. They have asked the people to brace up for challenging periods as they stabilize the macroeconomics they don't understand so that it can trickle down to the microeconomics they will appreciate. Unfortunately, the people of Africa have endured and hanged on these soothing but empty words for well over six decades now, but their circumstances have only moved from bad to worse. The other leaders are far worse. With guns and ammunitions, they cower the citizens into submission and do whatever is pleasing in their eyes.

Africans now find it difficult to believe in the economic terms that are meaningful only to the leaders but is of no relevance to the citizens whose conditions are worsening with each passing day. The political ideologies the leaders have tried to make the people understand, that it is the panacea to our troubles, is of no sense to us now because we now know. We know of capitalist countries whose health-care systems are solid, and we also know of socialist countries whose health-care systems are flawless. We know of kingdoms whose educational systems are impeccable, and we know of communist countries with superior educational standards. We know of countries ruled by so-called dictators with good infrastructure, and we know of democracies with sound infrastructure. It is only the African economics and political ideologies that put our health care, educational, and infrastructural systems in shambles when adopted. It is only the African economics and political ideologies that cannot feed her citizens and make them go hungry on arguably the most fertile and agro-conducive continent of the world.

"I am not an expert in economics, and I am not an expert in law. But I am an expert in working on an empty stomach while wondering when and where the next meal will come from. I know what it feels like going to bed with a headache for want of food in the stomach."[A5] These words were the response of Flt. Lt. Jerry John Rawlings to Major Okyere in 1979 as to why the former attempted an uprising in Ghana. If Africa's leadership will listen attentively, they will hear what the youth of Africa are saying—"Enough of your politics, enough of your economics, enough of your law, enough of your foreign degrees, enough of your expertise. Your accolades are irrelevant to our cause if all that you profess to have cannot provide us with the basics such as food for our survival."

But the people of African know that the negative statements made about the continent are wrong perceptions of our abilities and talents, but unfortunately, it is the picture we have painted over the years. Let the voices of our critics provoke our anger—a wave of constructive anger that is not directed to any other outside force; anger that is not directed at any person, country, or system but directed at ourselves. That we may see what is wrong on the inside and motivate us to move from our present state toward the dream of our founding fathers. In the words of Albert Einstein, "We cannot solve our problems with the same thinking we used when we created them." How can we expect to get a different result if we continue to do the same things in the same way?

Obviously, in the first few years of independence, Ghana set off on the right path in the right direction toward the right destination but is yet to reach her goal – economic freedom. The walk to economic freedom has taken longer than anticipated. It is imperative that we trace the paths that Ghana, the firstborn nation of black Africa, has treaded over the past sixty-two years. It is essential that we take a second look at the paths that brought Ghana and other African countries nowhere in over six decades. It might be useful to explore the curves, circles, U-turns, and winding paths that Ghana and in a broader sense Africa had taken and whether the end of the road is in sight. Maybe it will guide us back to the paths of our founding fathers.

A1   Myjoyonline, (2017), Full text: First independence speech by Kwame Nkrumah. Accessed December 1, 2018 through https://www.myjoyonline.com/news/2017/March-6th/full-text-first-independence-speech-by-kwame-nkrumah.php

A2   Abbey, R. A. (2018). The broken dream: 75% of Ghanaians want out of country – survey. Accessed November 22, 2018 through https://thebftonline.com/2018/business/the-broken-dream-75-of-ghanaians-want-out-of-country-survey/

A3   Nyabor, J. (2017). 1.7 million Ghanaians applied for US visa lottery in 2015. Accessed November 29, 2018 through http://citifmonline.com/2017/04/04/1-7-million-ghanaians-applied-for-us-visa-lottery-in-2015/

A4   Pew Research Center, (2018). At Least a Million Sub-Saharan Africans Moved to Europe Since 2010. Accessed November 22, 2018 through http://www.pewglobal.org/2018/03/22/at-least-a-million-sub-saharan-africans-moved-to-europe-since-2010/

A5   Happy Ghana (2018), Give me Ghana, Nigeria to recolonize; in a year they will become first-class countries - North Korean President. Accessed November 28, 2018 through https://www.happyghana.com/give-me-ghana-nigeria-to-recolonize-in-a-year-they-will-become-first-class-countries-north-korean-president/

A6   Heward Mills, D. (2011), Evangelism and Missions. Xulon Press.

CHAPTER TWO

Resource Waste: The Path Of Poverty

In the abundance of water, the fool is thirsty.
—Bob Marley

On July 10, 1953, Osagyefo Dr. Kwame Nkrumah tabled a motion before the people of England at the House of Commons, London. In that "motion of destiny," as it became popularly known, Nkrumah stated that "we [Africans] may lack those material comforts regarded as essential by the standards of the modern world because so much of our wealth is still locked up in our land."[B1] His statement was an acknowledgment of the apparent fact that Africans may be poor when measured by the skewed indices by which the modern world measures poverty and wealth. He was, however, hopeful that it will be just a matter of time that those same indices will declare Africans rich because the land of Africa is rich, and the precious and abundant resources of the nations will be ultimately transferred to the citizens of Africa.

Today, sixty-two years after leading Ghana to independence, with the fellow freedom fighters completing the liberation process he started, the people of Africa remain poor. Undoubtedly, something "bad" happened to the transfer of wealth process envisaged by Dr. Kwame Nkrumah, and the African of today is still looking for answers as to what happened.

The founders of Africa, by their speeches and actions, had hoped in the land of Africa. The hope of Africa was in their land and the riches that lie beneath it. The land was their hope because the dignity and pride of the African were shattered in colonialism, imperialism, slavery, exploitation, and degradation. The only thing that remained was the land, and, fortunately, the land was fertile with abundant precious resources. They anticipated that the status quo where the resources of Africa fed the industries of other continents, while Africans go hungry would change in independent Africa. The citizens of independent Africa were going to be fed by the proceeds of their resources, which feed the industries of other continents.

Six decades down the line, Africans are still hungry. Is Africa not extracting her gold? No! The noise of underground shafts in Johannesburg denies this. Is Africa not extracting her oil? No! The activities of multinationals at Nigeria's Niger Delta and the eighteen oil rigs scattered along the West African coast deny that fact. Is Africa no more mining her diamonds? No! The huge open pits of Angola's Catoca deny that. Is Africa no more extracting her steel? No! The sounds of the plants near Algeria's Oran deny that. Is Africa not felling her tress? No! The spaces in the previously dense and impenetrable tropical forest of Congo deny that. Has the soil of Africa refused to give its yield? No! The cocoa, banana, rubber, cotton, coffee, and other plantation fields of Sudan, Cote D'Ivoire, Kenya, and many other countries deny that. Does Africa lack market for her resources? No! The hooting ships on the shores of the Atlantic, Mediterranean, and Indian oceans deny that.

It is obvious that Africa has been extracting its natural resources since it became independent. The only question that is left unanswered is, where did the money go? Africa's resource wealth has not been transferred to her citizens. Where has it been moved to? To foreigners or the foreign accounts of Africa's ruling class?

THE EXPLOITATION OF AFRICA'S NATURAL RESOURCES

It will be difficult to find answers to the questions, but Kofi Annan gave a hint in the year 2013 that though "Africa's natural resources wealth rights belong to the continent's citizens, but these citizens are being robbed of its benefits by revenue diversion, corruption, jobless growth, and rising inequality." He added that "Africa and its partners will miss the opportunity to transform the lives of future as well as present generations if they carry on with business as usual [because] tax avoidance and opaque business practices block Africa's extractives sector.'[B2]

There are concerned Africans that are beginning to make their analysis and calculations to see where the monies intended for Africans have gone. One such group is Ghana's Centre for Natural Resources and Environmental Management (CNREM). The Executive Director of CNREM, Solomon Kwawukume in his numerous articles and petitions has raised a lot of issues regarding bad contracts in Ghana's extractive sector. In September 2018, the Centre for Natural Resources and Environmental Management (CNREM) published an article that provided an independent review of commercial oil production in Ghana after seven years. CNEREM's publication which is summarized in the tables below gave an idea about Ghana's oil resources which Ghanaians hailed in the year 2010 that the new oil resource was going to impact positively on the life of the ordinary Ghanaian.

Table A. Volume of Oil Extracted: 2010–2017

Oil Field	Barrels	Moving Average Prices US$	Value US$
Jubilee Field	221,860,786	83.4109	18,505,607,834
TEN	25,769,575	58.8998	1,517,822,813
Sankofa	5,455,512	45.5729	248,623,502
TOTAL	**253,085,873**		**20,272,054,149**

Table B. Allocation to Ghana

Oil Field	Barrels	US$
Jubilee	39,624,879	3,305,150,782
TEN	4,027,452	198,666,094
Sankofa	N/A	N/A
Subtotal	**43,652,331**	**3,503,816,876**
Taxes and royalties paid		523,304,893
Total Paid to Date to Ghana		4,027,121,769

Table C. Earnings by Contractors

	Barrels	US$
Contractor Parties Earnings	**209,433,542**	**16,244,932,380**

Table A shows that a total of 253,085,873 barrels of oil worth US$20,272,054,149 were extracted and exported by 31st December, 2017 from the three fields. Table B shows the total allocation to Ghanaians, the sovereign owners, was 43,652,331 barrels of oil worth US$3,503,816,876 representing 17.24% of the total production revenue. However, if Corporate taxes and royalties paid by the Contractor Parties per our findings were added, Ghana earned US$4,027,121,769, representing 19.86% of total production revenue from crude oil. The Contractor Parties earned 209,433,542 barrels worth US$16,244,932,380, representing 80.14% of total production revenue from crude oil. Gas proceeds earned by Ghana reported in the 4th Quarter of 2015 and 1st Quarter of 2016 Reports totaling US$9,856,621.

If Ghana had consolidated and adopted the Production Sharing Agreement (PSA) system, currently the world standard for equitable and fair fiscal regime which PNDC Law 84 introduced, Ghana would have lifted 151,851,523 barrels worth US$12,163,232,489 at little or no cost to Ghana. For not adopting Production Sharing Agreement, Ghana lost 108,199,192 barrels of oil worth US$8,136,110,720.[B3]

In the meantime, according to CNREM the Contractor Parties under-paid Ghana between US$800 million and US$902 million in Royalties and Corporate Taxes over the last six years under the Royalty Tax/Hybrid System.

The above investigations by a private civil society organization in Ghana point out one significant issue—the ability of the African leaders to negotiate for fairer deals for their people especially in her natural resources trade agreements. Many times, it is mind-boggling as one tries to understand certain agreements African governments get themselves into on behalf of her citizens. For some negotiations and contract, there are only two possible reasons to explain why such contracts are signed. Either those who negotiated (and in most cases African government officials) are dumb or they have been compromised to shortchange the country. No other reasons aside these two will be able to explain such transactions.

However, out of the two reasons, the latter seems to be the case. Most of the more significant resources contract African governments enter into are usually with the multinational companies. The multinationals are in most cases able to make "special satisfactory arrangements" with the few African government officials in such a way that the multinationals are given enough powers to do whatever they like including nonadherence to the country's environmental and other regulatory laws. In such cases, so far as the transaction is "satisfactory" to the government officials, it will be closed even if it is "unsatisfactory" to the entire country.

The country will be left at the mercy of the companies that give the country whatever percentage they like because those who negotiated the deal on behalf of the country are already satisfied with their deals and care less about whatever accrues to the state. The Multinationals seems comfortable with this arrangement because it is cheaper as it costs them less to satisfy a few senior officials than to satisfy an entire country. Dr. Peter Eigen, of Transparency International in an interview on Deutsche Welle (DW) Television wondered why the World Bank and Western Governments do not see anything wrong with the multinationals engaged in extractive industries paying tens of millions of dollars into private accounts overseas to secure bad agreements, contracts, and laws in their favor.[B4] More so, government officials ensure that that the multinationals have all the room they need to operate including silencing the few individuals or civil society groups who might raise questions and issues about bad deals.

In a Paper titled, "Concession to Service Contract," Ernest E. Smith expressed concern about the undue influences and corruption in contracts and called on Western Countries especially the United States to help curb exploitative behavior. Regarding the oil and mining industries, he stressed that; "some of these bad deals have already been producing resources, and the United States like other importing countries are consuming oil from some of these bad contracts. This places an important responsibility on the United States to lead by example in ensuring that oil and minerals from countries that promote questionable contracts tinted with corruption are not patronized."[B5]

The UN Security Council report on investigations into a conflict in the Democratic Republic of Congo found that government structures, army commanders, and businessmen were key actors in the "systematic and systemic exploitation" of the country's resources. From 1999 to 2002, the Kabila regime "transferred ownership of at least $5 billion of assets from the state-mining sector to private companies under its control in the past three years, with no compensation or benefit for the state treasury of the Democratic Republic of the Congo."[B6]

Also, according to a UN report, about one hundred people were killed in 2004 when Congolese military crushed a protest from civilians who were trying to demand more for the local workforce from an Australian mining company. Interestingly, according to the report, the mining company (Anvil mining) provided the logistics that sent the army to the mining village for the crackdown on the civilians. The company's chief executive found nothing wrong with making available the company's vehicles to the military because it was a "request from the army of a legitimate government."[B7]

Situations such as these will make it difficult for Africans to benefit from the wealth of Africa. It will be difficult for the continent to transfer its wealth to the citizens because of the "systematic and systemic" exploitation of the country's resources. The actors, as the UN report revealed, are many. It is a complex network of the African elites who seem to have a unique way of getting any democratic government officials into the net. The operations of the network are at times halted, but they are never stopped. While in opposition, a politician may have all the ideas and policies that can turn the fortunes of the country around; however, as soon as they get into government, the "elite network" will find a way to get the new government officials into the network one person at a time, and sooner or later, the "exploitation agenda" of the elite network is back on track. The exploitation agenda has been sustained for centuries, and it will demand something more from the citizens of the countries in Africa to break it.

The consequential effect of the exploitation is the tribal wars and other civil conflicts in Africa. As the elite keep impoverishing the people, the masses begin to lose hope in the government, and the youth especially become desperate. They become susceptible to anyone who will take up arms to fight the government. Civil wars then break out, and many people get killed. By the time peace is restored, the country will have to start over all again. Sooner or later, the exploitation will take another form, and the cycle goes on.

THE MACROECONOMICS OF AFRICA'S PRIMARY PRODUCTS

The issue of why Africa's natural resources have failed to lift Africans cannot be discussed without taking a look at the macroeconomics at play around the resources of Africa. The dynamics regarding the production and trading of Africa's resources are many, making it difficult to come out with a clear-cut solution for African countries. It is only when each African country makes an effort to take a comprehensive look at specific factors around specific resources will they be able to formulate and implement policies that will help in the growth and development process of the country.

Most of Africa's resources including tropical agricultural produce that Africa has comparative advantage producing like cocoa, coffee, cotton, etc., are not consumed or processed in Africa. These African products are sent mainly to industrialized countries who process them into products for local consumption and export back to Africa at much higher prices. Ideally, it is a good idea for African countries to set up industries to process some of Africa's primary product, especially for local consumption to save the country some foreign exchange on imports. The reality is that African governments have unsuccessfully tried this approach especially the founders of Africa whose import-substitution industries sprang up and died because they failed to have a look at the bigger picture.

Since Africa's commodities are sold in the international market, policymakers of African economies whose mainstay is in the production and export of primary products should have policies that are grounded on the behavior of the products on the international market where they are traded. Without understanding the dynamics, it will be hard to explain why Ghana, together with her neighbor Ivory Coast, produces 60 percent of the world's cocoa and yet has no say in the world market price of the commodity. In March 2018, even the slightest rumors of reduced yields from Ghana and Cote D'Ivoire sent cocoa prices to more than one-year highs within one week, but this power to move prices can't be used for the countries' advantage.

In publications on how commodity prices impact economies of developing countries, researchers Gersovitz and Paxson in 1990 and Deaton and Miller in 1995 give details of how demand, supply, and other factors play against African governments' efforts at obtaining inflows from exports to finance growth and development programs and projects. In his research work titled "Commodity Prices and Growth in Africa,"[B8] Deaton underscores that the long-run prices for most of Africa's commodities have remained the same for about a century with occasional ups and downs. He further posited that, though short-run events can lead to increases in commodity prices (at times for months), prices will eventually return to "base." The eventual return to base prices is driven by the fact that wages of the poor farmers cannot grow in the presence of unlimited supplies of labor at the subsistence wage.[B9] According to the researcher's theory, any improvements in the production process of the commodity accrue to consumers in industrialized countries and not the farmers and that, "in the long run, the market will complete the task that colonialism left unfinished."[B9]

According to the researcher, the forecast analysis of prices has inherent dangers. However, advisers to African governments, including the World Bank and IMF, often walk African countries along the misleading paths by recommending policies to solve misdiagnosed problems. For instance, in 1980, as the prices of copper began to tumble, the World Bank predicted that prices would bounce by 1985, which was never the case, causing serious revenue challenges to African economies that relied heavily on the commodity.

As Deaton suggested, "Import-substituting industrialization works no better when financed by commodity exports than by other means." African governments should be cautious about the seasonal price hikes of export commodities, which are often misdiagnosed as permanent revenues for the economies because the "real prices of primary commodities produced by African countries either have been without a trend or have trended gently down."[B9] The only way for Africa to increase the real prices of her commodities is to end poverty in Africa. From the short-term and long-term behavior of Africa's commodity prices, African countries that will devise and implement a broader strategy for achieving economic growth and development are more likely to be successful than total dependence on the export revenues of the country's commodities.

Fortunately, Africa continues to discover more and more hidden treasures. Currently, only five out of the fifty-four countries in Africa are not producing oil or looking for hydrocarbon activity. Ghana started producing the black gold—crude oil—in commercial quantities in 2010. New oil fields are currently being developed, which will increase production from the current two hundred thousand barrels per day to one million barrels per day in the short to medium term. After about four years of exploration, Kenya has discovered oil and gas deposits;[B10] Kenya's neighbor Tanzania has also found an additional 2.17 trillion cubic feet of possible natural gas deposits on shore.[B11] Nigeria's Ministry of Solid Mineral Development in the last quarter of 2018 announced the discovery of forty-four different minerals including gold, coal, copper, kaolin, limestone, lithium, platinum, granite, etc., in five hundred different locations across the country.[B12]

While still celebrating the discovery of oil in Uganda's Albertine belt, the country discovered other rare earth metals including aluminous clay, yttrium, gallium, and scandium, whose financial worth is estimated at $370 billion.[B13] Mali is set to produce lithium by 2020.[B14] The Rwanda Mines, Petroleum and Gas Board hinted of new deposits of minerals such as lithium, cobalt, iron, gemstones, etc. Mozambique,[B15] Zimbabwe, Cameroon, etc., are all coming out with new discoveries of hidden treasures as African countries step up exploratory activities in an effort to raise more revenue for development.

The primary issue of concern with these discoveries is that "on average, resource-rich countries have done even more poorly than countries without resources,"[B16] according to Joseph Stiglitz, a former chief economist at the World Bank. Discoveries of new resources will therefore not be enough to pull Africa out of poverty and underdevelopment. However, If African countries can make a better diagnosis of the real income from these abundant resources, and leadership can fight for a fair share of the resources from multinational companies, then together with a better developed human capital and reduced corruption, the wealth of African countries will in no time be transferred to her citizens. Failure to formulate and execute a broader strategy that will address the above challenges, Africa will continue on the paths to nowhere no matter how optimistic and wishful her citizens might be.

B1 Nkrumah, K. (1953) To-day we are here to claim this right to our independence, Motion of Destiny. Accessed December 1, 2018 through https://speakola.com/political/kwame-nkrumah-motion-of-destiny-independence-1953

B2 Ghanaweb, (2016). The modern conspiracy against Ghanaians. Accessed December 1, 2018 through https://www.ghanaweb.com/

GhanaHomePage/features/The-modern-conspiracy-against-Ghanaians-494097

B3 Kwawukume, S. (2018). Revised Seven Years Of Oil Production In Ghana -An Independent Report. Accessed December 1, 2018 through https://www.modernghana.com/news/879427/revised-seven-years-of-oil-production-in-ghana-an-independe.html

B4 Eigen, P. (2018, December 31). DW Journal Interview

B5 Smith, E. E. (1991). From concessions to service contracts. *Tulsa LJ*, *27*, 493.

B6 United Nations, (2002). Security Council is told peace in Democratic Republic of Congo needs solution of economic issues that contributed to conflict. Accessed December 1, 2018 through https://www.un.org/press/en/2002/sc7547.doc.htm

B7 United Nations (2004). United Nations Commission Mission to the Democratic Republic of Congo, Report on the conclusions of the special investigation concerning allegations of summary executions and other human rights violation perpetrated by the Armed Forces of the Democratic Republic of Congo (FARDC) in Kilwa (Katanga Province) on 15th October 2004). Accessed December 28, 2018 through https://www.ohchr.org/Documents/Countries/CD/LikofiReportOctober2014_en.pdf

B8 Deaton, A., & Miller, R. I. (1995). *International commodity prices, macroeconomic performance, and politics in Sub-Saharan Africa.* Princeton, NJ: International Finance Section, Department of Economics, Princeton University.

B9 Deaton, A. (1999). Commodity prices and growth in Africa. *Journal of Economic Perspectives*, *13*(3), 23-40.

B10 Sarkar, S. & Pema, T. (2017). Tullow says makes oil discovery in Kenyan well. Reuters. Accessed December 10, 2018 through https://www.reuters.com/article/us-tullow-exploration/tullow-says-makes-oil-discovery-in-kenyan-well-idUSKCN18D0SM

B11 Ng'wanakilala, F. (2017). Tanzania makes big onshore natural gas discovery - local newspapers. Reuters. Accessed December 10, 2018 through https://www.reuters.com/article/tanzania-gas/tanzania-makes-big-onshore-natural-gas-discovery-local-newspapers-idUSL8N16427G

B12 Vanguard, (2015). Nigeria discovers 44 mineral deposits in 500 locations. Accessed December 10, 2018 through https://www.vanguardngr.com/2015/11/nigeria-discovers-44-mineral-deposits-in-500-locations/

B13 Musisi, F. (2013). New minerals worth trillions discovered in Busoga region. https://www.monitor.co.ug/Business/Commodities/New-minerals-worth-trillions-discovered-in-Busoga-region/688610-2021266-by9acv/index.html

B14 Diallo, T. & Jabkhiro, (2018). Mali to produce lithium by 2020 with 694,000 T discovered. Accessed December 10, 2018 through https://www.reuters.com/article/mali-lithium/mali-to-produce-lithium-by-2020-with-694000-t-discovered-idUSL8N1XU2MF

B15. ANGOP (2017), Rwanda: New Minerals Found as Government Steps Up Exploration. Accessed November 30, 2018 through http://www.angop.ao/angola/en_us/noticias/africa/2017/1/7/Rwanda-New-Minerals-Found-Government-Steps-Exploration,3324b1b7-1391-4670-abe6-e8fa2dcf8de3.html

B16 tiglitz, J. E. (2012). From Resource Curse to Blessing. Project Syndicate. Accessed November 30, 2018 through https://www.project-syndicate.org/commentary/from-resource-curse-to-blessing-by-joseph-e--stiglitz?barrier=accesspaylog

CHAPTER THREE

Governance Institutions: The Path Of Disorder

Inspiring scenes of people taking the future of their countries into their own hands will ignite greater demands for good governance and political reform elsewhere in the world, including in Asia and in Africa.
—William Hague

No region of the world has there been constant calls for better governance than in the region of Africa. The Institute on Governance affirms that there is a need for governance anytime a group of people comes together to accomplish an end. The word "governance" came from the Latin verb *gubernare*, or more originally from the Greek word *kubernaein*, which means "to steer." Governance in the context of this book means steering the affairs of a country. In governing a state, there are so many things to direct and control, there are so many decisions to take, and there are so many things to do in such a way that setting up effective governance institutions is critical to ensuring that all the current and future needs of citizens are adequately catered for.

It is the responsibility of the government not only to create but also to maintain many central institutions. The broad term "institutions" includes government entities, laws, rules, and the informal rules of social interactions. Stronger institutions enhance the quality of governance, while weaker institutions do the opposite. Commenting on governance in Africa during his first official visit to Africa in July 2009, the forty-fourth president of the United States (U.S.) Barack Obama pointed out that "Africa doesn't need strongmen, it needs strong institutions."[C1] At the parliament house in Ghana where the U.S. president addressed the members of parliament and to the larger audience in Ghana and across Africa, the former president's words were seen by many as an attack on tyrants and dictators in Africa. Whatever be the interpretation, Obama's words undoubtedly refueled the lengthy discussion on why African continent is home for most of the world's longest-serving leaders who have been in power for more than two, three, or even four decades.

Across Africa, it is argued that the presidents usually orchestrate amendments to the national constitutions to extend terms of the president and that the ability of one person to plan and execute prolonged or even unending service terms for the nation puts the strength of governance institutions in question. Though there is nothing wrong with the extension of a president's term of office if it is carried out per the rules and laws of individual countries, the number of cases in Africa, however, raises questions. Chad, Algeria, Congo (Brazzaville), Togo, Cameroon, Gabon, Namibia, Tunisia, Guinea, and Burundi have all had terms of office extended for the president. Similar attempts in Zambia, Burkina Faso, Malawi Niger, and Nigeria, however, failed with accompanying violence and loss of lives.

To many Africans, Obama's call for stronger institutions in Africa was right, though many African voices had earlier made similar requests. Weaker institutions have cost Africa a lot and continue to drain the treasury of many African countries. The inability of institutions to curb corruption is estimated to cost Africa over $150 billion annually according to the African Union.[C2] The inability of African institutions to control and maintain internal and external security is creating conflicts that do not only wipe out economic gains but also cost countries billions of dollars to rebuild. Fake products including drugs that are meant to restore good health are also allowed into the countries of Africa because our institutions have failed to prevent them from entering our countries. The inability of our institutions to enforce the laws of the land has led to a blatant disregard for the law in such a way that, in some countries, only the fittest survives. Streams are polluted, forest are degraded, and streets are littered. Unwholesome products flood the markets, and criminals roam the streets while obedient citizens run for cover.

The call for stronger institutions seems to have fallen on deaf ears. Each country establishes institutions of governance to achieve specific objectives. There are so many institutions of interest, but the focus will be on just two universal institutions and how those institutions have impacted on governance in Africa.

The Military

There are only seven African countries that have not had an intervention from the military (Malawi, Botswana, Cape Verde, Eritrea, Namibia, South Africa, and Mauritius). African has experienced at least two hundred of both successful and failed coups since 1960, though many times the military stepped in at a time when things were moving from bad to worst. The military as a security institution is meant to protect the citizens of the state from external and internal aggression but not to govern. This is not to say that individuals within the military cannot be good leaders. Indeed, leadership training is an integral part of the military service. From recruitment to retirement, leadership is instilled more into officers of the military service than all the other security services, but to govern with that institution is to create chaos in the society.

In the twenty-first century, after the experiences of Africa with military interventions across the continent, African institutions should be strong enough to prevent the military from ceasing control of the state. There should not be any reason for the military to take control of the country if the institutions are working at least averagely. However, should there be the need for the military to step in to address any form of institutional failure, they must hand power back to a civilian government within the shortest possible time. From the continent's past experiences, any control of the state by the military over six months will mean doom for the country.

A major factor that works against military governments is the "legitimacy weakness" it takes upon itself as soon as it assumes the governance role. The people in a country should be able to take part in the political process of the country, and the rule of the military does not allow that broad-based participatory process. People look at military leaders who ceases power as illegitimate. To the people, the men in uniform are ruling over them just because they have guns and are skillful in the use of armory. Coercion, therefore, becomes the only weapon in governance for the military government, making it resort to the use of "kangaroo" courts, fear, intimidation, and force as the only way to administer the country. Under such conditions, it will be just a matter of time that another person will step in, and the cycle continues. Instability resulting from coups and countercoups have cost the countries in Africa a lot. From 1970 to 1982, there were seventy-two successful coups in sub-Saharan Africa.[C3]

In a globalized world, organized crimes provide a significant threat to countries without stronger institutions. Narcotics, drug trafficking, human trafficking, terrorism, etc., are fueled by weaker institutions of governance. Such crimes threaten the very existence of the people in a country. In Sudan, in particular, porous borders and underground economies, among other things, have made it possible to plan, finance, and execute all forms of organized crime. Somalia's porous borders among other factors made it easy for militants to carry out a terrorist attack in Kenya and escape according to a UN investigation.

The Judiciary

In Lesotho in 2018, Chief Justice Nthomeng Majara accused the government that it "seek[s] to remove me from office at all cost."[C4] Such reports raise issues of concern regarding good governance in Africa. The judiciary is one of the pillars of good governance. The ability of the judiciary to remain independent of other arms of government is critical to the relevance of the institution. However, just as recently happened in Lesotho, the independence of the judiciary in many African countries is being questioned. Reports of interference from other arms of government and reports of corruption in the judiciary are gradually eroding the confidence of the citizenry in the institution.

In a 2007 report by Transparency International titled "Global Corruption Report 2007: Corruption in Judicial Systems," corruption "is undermining judicial systems around the world, denying citizens access to justice and the basic human right to a fair and impartial trial."[C5] According to the organization's chair, Huguette Labelle, "Equal treatment before the law is a pillar of democratic societies. When courts are corrupted by greed or political expediency, the scales of justice are tipped, and ordinary people suffer," and "Judicial corruption means the voice of the innocent goes unheard, while the guilty act with impunity."[C5]

In Ghana, the investigative journalist Anas Aremeyaw Anas used just about a year to track some undercover dealings in the Ghanaian judiciary, and the results were bizarre. The secret recordings revealed high court judges were taking bribes equivalent to US$2,000 and releasing criminals on the street. Ghanaians have seen armed robbers taking $10,000 at gunpoint from victims. In other media reports, armed robbers have shot a victim and carried away over $100,000 from the street. The question that needs to be asked is that, if a judge takes $2,000 and releases a criminal after all the laborious work by police officers, state prosecutors, and individuals who came out with information to get an armed robber to face justice, what is the net profit of the criminal?

I will say, the net profit is $6,000. That is $10,000 less $2,000 for a judge, $1,000 for operational expenses, and $1,000 for the inconvenience of being detained for a few days in police custody. If this business is not profitable, then I am yet to see a lucrative deal. The above returns on the armed robber are calculated based on the worst-case scenario of being caught. If the robber is not caught, then the only expense to incur is the $1,000 dollar operational costs, which means the robber will be bolting away with a net profit of $9,000. If the judiciary should fail the citizenry in such a manner, a fundamental duty of the state, which is to protect its citizens, is destroyed. As a result, citizens will live in fear as lives and properties are destroyed by those who choose to earn incomes by illegitimate means. People who try to earn from legitimate sources become demoralized and find it risky to take loans for economic ventures because they are afraid of losing their monies to criminals at gunpoint. How long will countries allow her citizens to suffer by allowing criminals to thrive while good citizens continue to suffer?

AFRICA'S WEAK INSTITUTIONS

Several factors account for Africa's weak institutions. Some factors are internal and others are external, but the emphasis is on few internal factors discussed below.

Democracy Borne Out of Coups and Rebel Movements

African democracies have been impacted heavily by the military or rebel militia. In most African democracies, it is the military that supervised the crafting of the constitution that led the countries into democracy. The "Warlord Democrats," as the *Africa Now* magazine calls the phenomenon,[C6] participate in the elections as civilians and institutes democratic governance.

The 2002 constitution of the Republic of the Congo, which extended presidential terms from five to seven years, was crafted during the transition period under Denis Sassou Nguesso (a former military colonel), who also participated in the election and won by almost 90 percent of votes. After ceasing power in 1989, Sudan's president Omar Al-Bashir, in January 2010, resigned as a military officer to meet the legal requirements of contesting in Sudan's elections. The 1992 constitution of the Republic of Ghana was crafted under the military regime to Flt. Lt. Jerry John Rawlings, who resigned and participated in that same election and won.

The 2006 constitution of the DRC was crafted under the rule of Joseph Kabila, son of a former rebel leader who also participated in and won the election. Gnassingbé Eyadema rose to power in military coup d'état in 1967. In 1998, he organized presidential elections in Togo but canceled it when he was losing, citing "security reasons." His son Faure Eyadema was installed as president with the support of the army after his death in 2005. Faure went ahead to organize elections and won in 2005.[C7] Teodoro Obiang Nguema Mbasogo rose to power in a coup in August 1979. He has been the leader of Equatorial Guinea ever since. In the 2016 elections, the seventy-six-year-old president won another seven-year term as president with over 93 percent of the vote.[C8] Similar occurrences have happened in Uganda, Rwanda, Chad, Gambia, etc.

In such instances, the institutions of state are carved in such a way that they support the future aspirations of the would-be warlord democrat. In many cases, the president of the new democracy is allocated so many powers as to have the institutions of governance supporting him to have his way. In Ghana, for example, the president appoints almost every leader of the democratic institutions. The president appoints the Electoral Commission, the head of police, the head of the military, the head of the judiciary, the leaders of all security agencies, the auditor general, and the commissioner for human rights and administrative justice, among many others. These powers of "appointment" and "disappointment" make it possible for the president to control the institutions of the state in such a way that he can have his way even if he wants to rule for life.

Lack of Strong Opposition

In many African countries, it is "winner takes all" situation when it comes to winning elections and forming a government. Political parties are self-supported and receive little or no funding from the state. Also, the political parties are restricted from soliciting financing from other countries because foreign funds may come from illegitimate sources that might leave potential government officials in the hands of undesirable foreign operatives. Political parties that lose the elections, therefore, find it very difficult to fund their daily operations, limiting their ability to take an active part in the political process of the country.

Moreover, the winning party in an attempt to further weaken the opposition to perpetuate their dominance or rule most often use the institutions of state to harass and intimidate members of the opposition. As a result, many key opposition members are accused of inciting the public or other wrongdoings if they were previously in government. Members of the opposition are incarcerated or forced into exile, rendering them weak and unable to contribute fully in the democratic process.

Lack of a Strong Civil Society

Another factor that is contributing to the weak governance institutions in Africa is Africa's weakening civil society organization's participation in the democratic process. Analysts have observed a steady drop in the vibrancy of civil society in Africa, which has resulted in the inability of the civil society to exert and sustain pressure on Africa's political process. Nkwachukwu Orji believes "that the material bases of support for civil society organizations in African were eroded by the protracted economic crisis that gripped the continent from the early 1980s as well as the stringent neoliberal adjustment measures imposed with a view to resolving it. Consequently, many associations lost much of their organizational capacity to the extent that the state found them easy targets for co-optation or neutralization."[C9]

Gyimah-Boadi also notes that

> "the middle-class professionals and intellectuals who run key public institutions tend to be understandably preoccupied with their own economic survival, which often prevents their institutions from helping civil society to checkmate state hegemony. Judges depend on government for their appointments and for their operational budgets, and have few opportunities for lucrative private practice should they resign [or lose their job]. They can scarcely afford to maintain a posture of strict independence. Private newspapers fear losing much-needed revenue from government advertisement."[C10]

Poverty

Salaries and, by extension, income levels are unusually low in Africa. People would like to obtain other sources of income to supplement their incomes. However, most jobs in Africa, unlike in other developed countries, engage the worker from early morning to evening, leaving the worker with limited time to pursue other ventures that will help them earn extra income. The temptation to make extra income from the immediate environment (place of employment) is high. As a result, if the worker is unable to steal funds at the workplace, the tendency for the worker to take advantage of any opportunity that presents itself like taking bribe or engaging in corrupt practices at the workplace is high.

Another point that is associated with poverty is taxes and the high cost of securing government services. Whenever the cost of government service fees or taxes becomes too high or whenever the process becomes too bureaucratic, the temptation for officers of government to engage in corrupt acts becomes high. When citizens find taxes or charges to be too high, they prefer offering "lower bribes" to officers who can take them and give them what they want. Likewise, if the time taken to obtain government services becomes unnecessarily lengthy, people will want to pay something extra to cut down the waiting time.

On the issue of taxes, African countries as developing economies need a lot of money to fund her infrastructure. Day in and day out, new taxes are imposed by governments in an attempt to shove up the country's revenues. Governments have resorted to increasing taxes and costs of government services. This is, however, being counterproductive.

For instance, when businessmen in Ghana complained about the high cost of duties and taxes associated with imports and the government failed to address the concern, the situation became fertile grounds for corruption for customs officials at the various points of entry into the country who cashed in on bribes. The former president Atta-Mills, on paying a surprise visit to the Tema Port after an undercover investigation revealed customs officials taking huge sums of money from businesspeople at the expense of the state, could not hide his feelings. He lamented over the revenue leakages and the involvement of high-level customs officials who connived to loot the country. To the president, he now understood why people were eager even to pay bribes to get into the customs service because he heard that if one person becomes a customs official, it takes less than a year for that individual to raise mansions and drive new and better cars. He heard it but didn't believe, but now he believes. He questioned if an officer's salary can build him a house after even ten years of service. He vowed to weed the bad ones out, but he never lived to accomplish this mission.

Recruitment Process into Governance Institutions

The institutions of governance should have the best quality human resource because leadership is a crucial factor in the success of any venture. The actions of institutions of the state affect all people and organizations in the country and, as a result, should have the best personnel. However, the high rates of unemployment in many African states are having an impact on the recruitment process into the various institutions of governance. Currently, it takes something extra to gain access to state institutions. Recommendations from a person in the high positions are helpful when seeking appointment into the police, military, immigration, customs, and other various state agencies.

Because of the high rate of unemployment, people that do not have the desire to be in the service to serve their country want to be employed because it is seen as a means of obtaining income. Officers with no passion or love for the service are being recruited into the service because they have the "connections" to get them recruited into the service. As noted by Chabal and Daloz (1999) and Lindberg (2003), "It is often through political connections that people gain access to contracts, employment, education, and land. Elections are thus not only an issue of which policy direction a state should choose but also about everyday economics that can have a decisive impact on citizens' well-being and even survival."[C11]

Technology to Support the Institutions

The inability of African countries to adopt appropriate technologies to enhance efficiency and effectiveness in state institutions also affects the strength of the state institutions. State institutions, by their nature, deal with a lot of data. Advanced information management systems are needed to make the institutions efficient and for them to serve their purpose. More information on the technology is provided in the ensuing chapter.

BUILDING BETTER GOVERNANCE INSTITUTIONS

There is no quick fix to strengthen Africa's weak governance institutions because of the multiplicity of factors. That is not to say it cannot be done. It can be fixed by adopting the right strategies. A significant step that can lead to the strengthening of Africa's institutions is for Africa to tackle poverty levels and utilize technologies to enhance governance and check abuse.

Besides, African countries should be able to prove their case in seeking development assistance from other countries or other international organizations. Africa has to learn her lessons. Suggestions from well-meaning countries and institutions should be considered and refined for adaptability within the context of Africa. Approval from other countries doesn't mean Africa is on the right track. Setting up well-defined goals and evaluating success based on the goals are the best ways for Africa to better her governance.

The pressure on African governments from outside Africa to raise revenues may be from good intent. Indeed, Africa should buck up regarding revenue mobilization as it is needed to support the diverse and growing needs of the populace. But there are several ways to achieving a single goal. Though some may be costly than others and some might take a more extended gestation period than others, the core objective or target remains the same. It behooves on African governments to choose whichever way it wants to go in achieving set goals and be able to defend that course of action when it matters most. Failure to offer alternative actions leaves the government with only one choice—to "swallow" whatever is proposed by the lenders.

If in business transactions individuals can convince financial institutions to alter lending terms and conditions to suit a particular business that one intends to operate, there can be no reason why governments will not be able to convince their creditors to alter the terms and conditions of a loan agreement. The only obstacle envisaged is when the government does not know what it is going to do with the money. If the governments mean business and are willing to prove a point, just as Kwame Nkrumah wanted to demonstrate the capability of the black man in governance, then governments should be able to change the position of their creditors regarding amount, repayment plan, and other terms associated with the loans and grants agreements.

A case in point is when it comes to mobilizing revenue. Not all increases in taxes lead to an increase in government revenue. Sometimes, reductions in taxes lead to increases in government revenues. At times, the increases in taxes create more incentives for people to go behind the system to offer bribes and corrupt the system. In the end, the state, in an attempt to raise revenues by raising taxes, receives far lesser than the income it was previously receiving. Africa needs to take a second look at its large informal sector and how to formalize the various informal sectors. The introduction of value added tax (VAT) and other forms of taxes is not sufficient to bring in the desired revenue from the informal sector. VAT, for example, imposes double taxation for formal workers, while it taxes the informal worker only once. For example, the teacher in the formal sector will buy a tin of milk and pay VAT, and the plumber who does not work with a registered company (informal sector) also buys a tin of milk, which has the same VAT component. What the two have in common is the VAT they all pay when accessing goods and services, but the teacher pays an "extra" income tax, while the "freelance" plumber walks away with no tax burden. Certifying artisans in the informal sector will not only ensure quality services but also bring them to a form of formal structure that will be more receptive or responsive to government policies.

Related to the issue of raising government revenue is providing adequate funding to governance institutions. Aside from maintenance, institutions need to grow and evolve to meet the ever-changing needs of the society to function at their optimum level. This objective cannot be achieved without funding to support its growing process. It is therefore incumbent on African governments to continue to provide governance institutions with the financing needed for operations, research, and development. Many government agencies exist just in name and cannot even afford to obtain basic things for office operations. There is the need for major reforms in most African institutions of governance, especially the security and judiciary forces. This cannot be achieved without the needed funding.

THE ATTITUDE OF THE GOVERNED

It will not be a proper touch on the government without touching on the governed. One of the lecturers of my MBA course that has since been on my mind is that of a marketing lecturer who said, "In the river of marketing, you don't swim opposite to the river's current. Always swim in the direction of the river because it is suicidal for the firm to swim against the river's current as it will destroy the finances of the firm which is the backbone of the firm." He continued. "Giving to the people what they want is like swimming with the river's current, it is easy because the resistance is weak and the river's energy supports your efforts. Giving to the people what they want is flowing with the river's current but trying to give them what you think is the best for them is swimming against the direction of the current. Knowing the tastes and preferences of your customers is critical to successful marketing." These words are fresh in my mind today, though it has been over a decade.

These words prompt me to take a closer look at leadership and followership; government and the governed in Africa. Democracy is widespread in Africa. Almost every country in Africa is now democratically governed, though the democratic process that got them elected may have questionable democratic features. In a democracy, power is given to the people to choose their leaders. An essential aspect of democratic governance is how people decide whom they want to lead them in an election. Sociological factors such as gender, education, age, religion, occupation, background, income, etc., influence voters to vote in a particular way. It appears, however, that psychological factors such as the candidate, political party, and critical policy issues are key determinant issues that voters consider when going to the polls. It means that what a political party stands for and key policy issues that are put forward during the campaign period are major determinants in choosing leadership in Africa provided the democratic process was fair.

When Africans are given the opportunity to go to the polls, they elect the kind of leaders that they want. The question is, what policy issues or message does these politicians give to the electorate to get to their votes? In context with my marketing lecturer's advice of flowing in the direction of the river's current, I project that politicians in Africa, in their campaigns, tell the voters what they want to hear and do not try to provide what is right for Africa. Politicians in Africa give to the people in Africa the messages that they want to hear. Politicians avoid telling the people of Africa the sacrifices the people will have to make to support the development of their countries, which will elicit resistance and will not get them the votes they need for political power. They instead tell the electorate what they want to hear. Messages such as the government will pay for your housing, your health-care bills, your kids' school fees, etc., are the most popular messages among the electorates. Policies that are not well thought through easily fly on campaign rallies and elicit thunderous responses from the voters. In a nutshell, Africa is getting the leadership that the people want.

Is the African ready to receive the truth? Is the African electorate prepared to accept the truth? Will the African electorate vote for the person or the politician that will tell them the truth? They are people with itchy ears that want to hear great promises—promises that will be difficult for them to fulfill. Politics has been reduced to who can lie best. Who will be able to tell the lie that will meet the needs and the yearnings of the people? Throughout the constitutional history of the African continent, presidents have ridden to power on promises that they have failed to fulfill, but Africans have not learned the lesson. The politicians promised that they were going to bridge income gap so the people who live in the rural communities, the poor, will be able to have the same access to the facilities that are being enjoyed by the rich. Politicians who have come that they are going to get the unemployed employed. Some people have come with the promise that they are not going to take more taxes from the people. In the end, after receiving the legitimacy or the mandate of the people, they failed to deliver what they promised.

Is it the people who are calling for the politicians to keep on lying to them? Because if the only way to get the vote of the public is to lie to them, then why would you tell the truth and lose? Why would you swim in the opposite direction of the current if you want to market yourself to the people? Why would you tell the truth to struggle to get elected if you have the easy option telling the people what they are ready to listen to?

The African needs education, better health care, employment, housing, and all basic human needs. The politician knows that the

African is in dire need of such things, but poverty is affecting their ability to have access to such things. As a result, one way of getting the electorate votes is to market a candidate that will be able to provide such things at no cost to the electorate. This marketing strategy is delivering power to the voters but plunging Africa into serious debts. A key contributor to Africa's growing debts levels has its roots in such promises. This is because, as governments seek reelection, they are somehow challenged to redeem promises, and this forces them to seek funding to honor their commitments. In such instances, their priority does not lie on how much debt they are accumulating or the repercussions on the future of the country. Their priority instead is how to redeem the promise and retain power. Many African economies are heaping up debts on the unborn generations of Africa because they want to satisfy a current need without a sustainable strategy.

For instance, in Ghana, the NPP promised and provided free health care, though it came as a cost to the ordinary Ghanaian as VAT was increased by 2.5 percent to support National Health Insurance Scheme. The NDC also provided free school uniform and exercise books for primary schools amidst complains of too much taxes from the populace. The NPP again promised free senior high school and though delivered on it, the country is still struggling with funding and infrastructure to sustain the policy.

If Africa needs free education, the people have to bear in mind that somebody has to teach the lessons and that person has to be paid. If Africa desires free clothing from her politicians, the people should bear in mind that somebody has to be paid to sew the clothing, and someone has to make the clothing material. If Africans desire free food, the people have to appreciate the fact that there are farmers who sweat to produce agricultural goods, and they have to be paid. So when a government promises free things, the interest of the African should not be what is being promised but rather how the promises are going to be delivered without creating another problem that may even be bigger than what they are trying to solve.

What Africa needs are governments that are committed to mobilizing revenue without putting excessive pressure on its people. Governments committed to making sure that the proceeds of the resources that the country has are not eroded by corruption. Governments that will ensure that it gets the best out of the resources that the country is selling to the outside world. Governments that will open up the economies of their respective countries in such a way that they can empower the citizens to work so that they can tax them to provide for their social needs. Let the electorate know this so that they can ask the hard questions when the politician begins to make promises.

C1 Baker, P. (2009). Obama Delivers Call for Change to a Rapt Africa. New York Times. Accessed December 15, 2018 through https://www.nytimes.com/2009/07/12/world/africa/12prexy.html
C2 Hanson, S. (2009). Corruption in Sub-Saharan Africa. Accessed December 15, 2018 through https://www.cfr.org/backgrounder/corruption-sub-saharan-africa
C3 McGowan, P. J. (2003). African military coups d'état, 1956-2001: Frequency, trends and distribution. The Journal of Modern African Studies. 41. 339 - 370. 10.1017/S0022278X0300435X.
C4 Rickard. C. (2018). Lesotho: Independence of the Judiciary in Peril. Accessed December 15, 2018 through https://africanlii.org/article/20180518/lesotho-independence-judiciary-peril
C5 Transparency International (2007). Global Corruption Report 2007: Corruption in Judicial Systems. Cambridge University Press.
C6 Themnér, A. (Ed.). (2017). *Warlord democrats in Africa: ex-military leaders and electoral politics.* Zed Books Ltd..
C7 Banjo, A. (2008). THE POLITICS OF SUCCESSION CRISIS IN WEST AFRICA: THE CASE OF TOGO. *International Journal on World Peace, 25*(2), 33-55. Retrieved from http://www.jstor.org/stable/20752832
C8 Wikipedia (2016). 2016 Equatorial Guinean presidential election. Accessed December 15, 2018 through https://en.wikipedia.org/wiki/2016_Equatorial_Guinean_presidential_election
C9 Orji, N. (2009). CIVIL SOCIETY, DEMOCRACY AND GOOD GOVERNANCE IN AFRICA. CEU Political Science Journal, 4(1).
C10 Gyimah-Boadi, E. (1996). Civil society in Africa. *Journal of Democracy*, 7(2), 118-132.
C11 Chabal, P., & Daloz, J. P. (1999). Africa works: Disorder as political instrument (African Issues). *James Currey, Oxford.*

CHAPTER FOUR

Lawlessness: The Bumpy Path

Laws made by common consent must not be trampled on by individuals.
—George Washington

When I first heard about the World Justice Project (WJP) a few years ago, I was particularly curious as an African to see how Africa will perform in the rankings in subsequent years, but things did not change that much especially in the direction I wanted it to go. The WJP is an independent multidisciplinary organization that works to promote the rule of law worldwide. WJP produces the annual Rule of Law Index (RLI) that measures a country's adherence to the rule of law. In other words, the index measures the extent of lawfulness or lawlessness of a state. World Justice Project recently released the 2017–2018 rule of law index where African countries constituted 50 percent of the ten most lawless countries of the world but constituted 0 percent of the world's most lawful countries. Widening the list further to the top and bottom-twenty countries, African countries represented 0 percent of the twenty most lawful countries but 45 percent of the twenty most lawless countries. No African country is among the world's forty lawful countries.

RANK	COUNTRY	RULE OF LAW INDEX
1	Denmark	0.89
2	Norway	0.89
3	Finland	0.87
4	Sweden	0.86
5	Netherlands	0.85
6	Germany	0.83
7	New Zealand	0.83
8	Austria	0.81
9	Canada	0.81
10	Australia	0.81
11	United Kingdom	0.81
12	Estonia	0.80
13	Singapore	0.80
14	Japan	0.79
15	Belgium	0.77
16	Hong Kong SAR, China	0.77
17	Czech Republic	0.74
18	France	0.74
19	USA	0.73
20	Korea	0.72

The twenty most lawful countries

RANK	COUNTRY	RULE OF LAW INDEX
94	Liberia	0.45
95	Kenya	0.45
96	Guatemala	0.44
97	Nigeria	0.44
98	Madagascar	0.44
99	Nicaragua	0.43
100	Myanmar	0.42
101	Turkey	0.42
102	Bangladesh	0.41
103	Honduras	0.40
104	Uganda	0.40
105	Pakistan	0.39
106	Bolivia	0.38
107	Ethiopia	0.38
108	Zimbabwe	0.37
109	Cameroon	0.37
110	Egypt	0.36
111	Afghanistan	0.34
112	Cambodia	0.32
113	Venezuela	0.28

The twenty most lawless countries

Source: WJP Rule of Law Index 2017–2018 (World Justice Project)

From the statistics, lawlessness in Africa is high. For those who have had the privilege of traveling across the continent, a survey is not needed to show that lawlessness is pervasive on the continent.

There seems to be a correlation between lawlessness and development. Undoubtedly, laws and its enforcement thereof is central to Africa's development efforts. It is apparent from the Rule of Law Index (RLI) and the Human Development Index (HDI) of 2017 that the top-ten lawless countries have lower HDI, and vice versa. Achieving a group goal is quite more difficult than realizing personal objective. The reason is that for a shared objective to be achieved, it requires that all the members commit themselves to behaving in a certain way, as the behavior of each member in the group affects the group's efforts. Managers of group efforts always ensure that those who exhibit inappropriate behaviors are "corrected" or inhibited. The success of the Ancient Greek city-states hinged on this basic principle.

Ancient Greek states long before the birth of Christ had laws governing human behavior that are similar to the laws of modern sovereign states. Such laws were so basic that the Greek philosophers referred to them as natural laws, universal laws, or the laws of nature. Aristotle referred to natural laws as "the unwritten but universally recognized principles of morality," and it is "reason unaffected by desire."[D1] Humans by nature have desires—the desire to control and to have or acquire material things, the desire to seek pleasure, and the desire to feel loved or valued and to know or understand, among many others. The forces behind such desires are often so powerful that they at times override or take over the human mind's function of controlling actions. If that desire is an evil desire, it tends to unleash misery on humanity, quite often the victim and the perpetrator.

The end product of evil desires is most often devastating. In 2017, a blogger Ivan Šimonović wrote the following about the Central African Republic in an article titled "Lawlessness in the Heart of Africa":

> The courtroom in Bambari town has broken doors and no windows. The top of the judges' bench, tables and chairs are all gone—looted. The floors

are strewn with papers—the remnants of archives and civil registries. The justice system in the Central African Republic today looks exactly like this courtroom. Outside the capital, Bangui, there are no police, no prosecutors or judges, but Séléka coalition forces, who took control of the country in March, are everywhere. They receive no salary; instead, they set up checkpoints or go to the market and to extort money, or loot houses. But they do not just loot. In Bambari hospital, I met a number of their victims. Eighteen-year-old Martine was four months pregnant when she was raped and had a miscarriage. Solange, a widow, and mother of five was raped as she sold coffee by the road. Annette was shot in the arm because she would not let go of the little money she had earned at the market. Many have been killed, but no one knows the exact numbers.[D2]

Many countries of Africa find themselves walking on paths similar to the one described above. A total breakdown of law and order resulting from conflicts, civil unrest, the military takeover of governments, disputed elections, desperate attempts by people whose desire is to hold on to power, and many more.

These happenings do not only bring countries of Africa to their knees but also erode the gains, destroy key human resources, and send many as refugees to other countries who might never return to rebuild their families let alone their country. The destructive nature of such occurrences is apparent, yet some African countries continue to have repeated doses of such events. Military coups are uncountable; civil wars have hit Nigeria, Chad, Mozambique, Cote D'Ivoire, Liberia, Mali, Niger, Mauritania, etc., more than once. Violent election disputes have hit Kenya, Zimbabwe, Ethiopia, Nigeria, Gabon, Gambia, and several other African countries. Such events are uneven paths that have slowed down Africa's economic progress.

Some African countries have been able to avoid the paths of conflicts and violence—at least for a decade. It is expected that such countries move closer to their economic goal, but they also have their own challenges—lawlessness resulting from the inability to enforce existing laws. These countries are just eager to make rules, but there are no conscious efforts to enforce the laws. There is no order as people do whatever is pleasing in their own eyes. Drivers drive at 120 km speed on roads marked with 50 KPH at the full glare of the police. People throw garbage anywhere they find appropriate. People ease themselves on beaches and anyplace they find appropriate. Pickpocketing has become norm, and it has become the responsibility of the individual to watch their pockets, while pickpockets are at liberty to steal from those who lose their guard. There are ad hoc government regulations such as restricting press freedom and bribery. There are extortion, intimidation, and brutalities from law enforcement officers and public officials. The laws have become discriminatory, allowing certain groups of people, if caught, to go free, while others, especially the poor, have to face the punitive measures.

Lawlessness in Africa is mainly fueled by the interventions from "higher places" when a person commits a crime. As a result, whenever law enforcement officers see someone disobeying the law, they are reluctant to act because they don't know the background of the individual or how "connected" the person might be. The officers, instead of focusing on enforcing the law, now focus on how the "enforcement action" will be unproductive or even negatively affect their career. Whenever an enforcement officer in many African countries sees someone breaking the law, three different scenarios play in the mind of the enforcement officer instead of one. The first scenario is what I call the "ideal" scenario, which is to take the necessary actions to halt the undesired behavior of the perpetrator and bring the perpetrator to justice under the laws of the land. The "ideal" scenario is what the law enforcement officer is trained to do. This is what happens in lawful countries.

The second and third scenarios characteristic of many African countries are what I call the "wasted-efforts" scenario and the "career-conscious" scenario. In the "wasted-efforts" scenario, the officer reflects on the question, will this lawless person face justice or is capable of using money or connections to people in "high places" to block justice? Here, the officer uses personal judgments and will decide whether the efforts he will exert as a law enforcement officer will yield positive results. If he is persuaded that his effort is not going to produce any fruitful results, he will let go of the lawless behavior and mind his own business. If he is convinced the individual will face the law, then he will take the "ideal" scenario action.

In the "career-conscious" scenario, the officer asks himself the question, *Is this person engaging in illegal conduct "connected" to any person influential enough to punish my legal actions?* Here too the officer uses his personal judgments to check if exercising his primary duty will put his career in danger, and if it will, he will not take any action. This means misery for the poor and the ordinary citizens of the land. They are always in the law's net, and they face the full rigors of the law when they err, but that cannot be said of the rich, famous, and "connected." The law has been turned upside down. Those who are planning and executing evil are becoming bolder by the day, and those who are trained and paid to keep people and property safe are becoming fearful by the day. Many law enforcement officers are unable to act because of the fear that doing what is right can lead them into trouble.

The disturbing part of the issue is that the criteria used by the officers to judge whether the lawless person will face the law are ridiculously empty. Usually, it is how the person carries himself or how the person is dressed or the car he may be using at the time of the act. These are physical judgments they make in a couple of seconds. I recall an incident that happened at a road toll booth in Ghana. The car we were traveling in was behind a black Toyota Prado as we approached the toll booth. When the Prado got to the toll attendant, I was expecting that the windows will come down for the driver to pay the toll. To my surprise, the driver didn't bring down the windows, and he didn't pay but instead sped off. I heard the lady in the booth shout. Then the police officer who happened to be sitting behind the booth stood up with his gun in his right hand and just gazed at the vehicle. I saw a police motor, which I believed belonged to the policeman parked close to him, but he did not make any attempt to chase the car. The police officer sat down as we all watched the car disappearing in the horizon. It was our turn to pay, so the driver driving our car asked the lady if the vehicle ahead of us paid. She replied in the negative. So the driver said, "If some of us have to pay and others will not pay, then, next time, we are all not going to pay."

The point I want to make from this incident is that officers of the law play all the three scenarios in their mind before they take any action to stop a wrong behavior. The "ideal" scenario is what will play first because they have been trained to do that, and they will act in many times unconsciously. That is what prompted the policeman at the toll booth to jump on his feet with his gun in his hand at the sound of the lady in the booth. That impulse to act ideally is then challenged by the other scenarios: First, the "wasted-effort" scenario; "if I pursue this man and arrest him, can't he stop the prosecution process with his money or power?" Additionally, "career" scenario also comes; "is he not a relative of an influential politician that can cause my demotion or dismissal?" And looking at the Toyota Prado, he becomes convinced that the person is rich enough to either pay his way out the law or can even cause his transfer or make his life uncomfortable in the service. The best action to take then is to allow the car to speed off for the state to lose money so he can save his energies and probably his job.

The inconsistent paths where the laws of Africa deal ruthlessly and mercilessly with the poor and the ordinary citizens of the land but sympathetically and mercifully with the "rich" and "influential" citizen are a dangerous and slippery path. It only encourages people to take the law into their own hands. As declared by former U.S. president John F. Kennedy, "If a free society cannot help the many who are poor, it cannot save the few who are rich."[D3] Every person can make use of personal protective and survival mechanism because even tiny animals do, but for uniformity and order, the state has laws and hires law enforcement officers to enforce the laws. The discriminatory application of the law is a primary contributor to chaotic situations we find in most African countries.

I observed a different incident to the toll booth incident in the United States a few years ago. Kojo, a friend of mine, had to drive me from Woodbridge, Virginia, to New York City, New York, to catch a flight back to Ghana. The GPS estimated that it would take us five hours, but Kojo told me it had factored in the traffic and freeway limits, so with a bit of luck in the traffic situation and a few miles per hour over the speed limits, we should be able to catch the flight.

On the road, I observed he kept looking over his shoulders and into the rear mirror.

When I asked why, he told me that he was over the speed limit and the police are on the road. He said the police drive in unmarked vehicles, so it is difficult to tell which car is a police car, so he ought to be careful because he can't lose his license to drive. I told him that he had a genuine reason to speed up, and since I have my passport and ticket, they will reason with him for a few miles over the speed limit. I was shocked when he told me that "catching a flight is not an excuse to put your life and the life of other road users in danger." I said to myself, how can a Ghanaian be such law-abiding in the United States but chose to do whatever he wants when he is in Ghana?

In Africa, people give excuses for anything including reasons for a blatant disregard for the law. Africans give excuses for being late at work. Excuses for absenteeism, excuses for nonperformance, excuses for driving over the speed limit, excuses for overloading, excuses for not paying taxes, excuses for not sending their children to school, excuses for building in waterways, excuses for neglecting their children, and excuses for whatever you can think of. It is not a surprise that African countries have excuses for being underdeveloped after decades of self-governance. Until the African continent runs short of excuses for her failures and obtains a reason for achieving economic success, real economic development will continue to elude the continent. Africa was under colonialism for many years until some Africans—the freedom fighters—found no excuse to be under the control of foreign governments. Africa will be ready for economic success when a higher percentage of her people see no reason for the present economic state.

Back on Virginia to New York freeway. I observed the signage well and told Kojo that the speed limits are not all that visible and he could just say that he didn't see the speed limit. His response was that my idea will put him into bigger trouble.

"I prefer to remain silent if I am caught than to make such a reckless statement," he said.

"Reckless?" I asked surprisingly.

"Yes, reckless because that will imply that I don't look out for road signs when I am driving," he replied.

According to Kojo, that will be a severe breach of the law, and his license can be revoked because it is the responsibility of the driver to read and adhere to every sign while on the road.

Kojo is a changed man! This is not the Kojo I knew in Ghana. He didn't talk like this, and he didn't behave like this. He didn't reason like this, and he was the worst driver I knew—a habitual road traffic offender who threw verbal attacks on the Ghanaian police and dared them because he was a wealthy and connected businessman. The police personnel in "higher offices" were his friends, and that was his license to flout the country's traffic laws. I don't know any business laws he flouted, maybe several of them, but I can only guess it from the way he flouted the traffic laws. He will extend that to other circles of operation because he loves to stay connected, and he spends a lot of money on people in authority so he could use them when he is in the arms of the law. My surprise is how Kojo has become a tamed driver after just two years in the United States.

I missed the flight, and I had to pay $300 more to fly the next day. But what matters most is that I am alive and well. Thanks to the better motor traffic system of the United States. Anything could have happened if Kojo could drive in the United States the way he does in Ghana, but the good system made sure it didn't happen. I made it to Ghana safely.

According to James Allen, "A man does not live until he begins to discipline himself; he merely exists."[D4] In the perspective of James Allen, a more significant percentage of Africans merely exist because they exhibit unruly behaviors or are undisciplined. There is the urgent need to find a way to instill discipline among citizens of African countries. Most African countries have embraced democracy. Democracy means the rule of law. The law reigns, and everyone falls under the law. But in African democracies, some people reign with the law, and such people have the opportunity to twist the rules in their favor. People find it difficult to sacrifice their freedoms in respect of the law for the effective functioning of society.

The status quo regarding lawlessness in Africa is disturbing. While the rich and the political class easily get away from the long arms of the law, the ropes get tighter for the poor, and the opportunities for the poor to get justice are decreasing by the day. It takes years to have a verdict on a seemingly simple civil case. The slow justice procedure with the unnecessarily frequent adjournments coupled with the high costs of accessing the services a lawyer is a disincentive for the poor to obtain justice from the courts.

As proposed under the chapter "Governance Institutions," the justice system in Africa needs major reforms. The reforms should be able to shorten the time for seeking justice. It should eliminate unnecessary procedures and irrelevant colonial structures. The changes should make the justice system more accessible to the poor. The reforms should employ modern technologies to curb potential bribery and corruption. The improvements should further strengthen the independence of the judiciary and disentangle the judiciary from any manipulative cords of the executive. With a stronger justice system, the law will take its rightful position above every individual and individual desires establishing the order required for economic success.

Political leaders in Africa should also show leadership. Many people use their offices or "connections" to high offices to intimidate law enforcement officers and to escape from the law. Some of such intimidator acts occur at the blind side of the officers holding power. Pres. John Atta-Mills, a former president of Ghana, showed leadership when he got to know that unscrupulous people were using his name or office for their gains and made the following remarks to the customs and police officers at the Tema Port in the presence of the press:

"There are people who come here [seaports] throwing their weight around with the names of so-called 'high ups.' I have told General Morgan, and I'm telling you now. If anybody comes with the president's name, the first thing you should do is to arrest that person."[D5]

With these words, the former president showed the way; he was leading by example. He proved that he had the political will to pursue the economic objectives of his country in three ways; First, the statement removed the fear of the president's office or higher offices from the minds of the law enforcement officers, making it easier for them to exercise the law of the country without fear or favor. It meant that officers are at liberty to apply the law without the fear that their career may be at risk or their efforts will not amount to anything. Second, the president was saying he will not be bothered even if his officers are arrested, detained, or sent back empty-handed, if only that is what the law says. Thus, he was willing to cooperate with the security institutions to have justice served on any person who falls foul to the law. Third, he showed leadership to the people holding high offices in the land that, if even on the authority of the first gentleman of the land, nobody can break the law and not face justice; they should not expect that they or their associates can break the law and get away with it. This is the kind of leadership that is certain to set countries on the paths of economic success.

D1 Edmunds, S. E. (1924). The lawless law of nations. . *Louis L. Rev., 10*, 171.
D2 Šimonović, I. (2017). Lawlessness in the Heart of Africa. Accessed December 15, 2018 through https://www.huffingtonpost.com/ivan-simonovic/lawlessness-in-the-heart-_b_3755302.html
D3 Brainy Quotes, (2017). John F. Kennedy Quotes. Accessed December 15, 2018 through https://www.brainyquote.com/quotes/john_f_kennedy_125480
D4 Allen, J. (2012). *Above Life's Turmoil: eBook Edition.* Jazzybee Verlag.
D5 GWS Online (2018). President John Evans Atta-Mills storms Tema Port. Accessed December 15, 2018 through https://www.ghanawebsolutions.com/videos.php?v=ed57WhR5ouQ

CHAPTER FIVE

Dependency Mind-Set: The Path Of Stagnation

A dream doesn't become reality through magic; it takes sweat, determination and hard work.
—Colin Powell

African countries have shown signs that it might not be able to be economically independent. One visible sign is that, after six decades of political independence, Africa is nowhere close to being economically independent. Though many have attributed Africa's slow progress to factors such as colonialism, brain drain, exploitation of Africa's resources, among others, that argument may not be plausible because other economies like Israel have gone through similar experiences but have been able to overcome such challenges and placed themselves as strong global economies. It appears African countries and, to a greater extent African leadership, have over the years developed a dependency mind-set that is going to keep Africa dependent on other nations for an extended period.

Only a few will doubt that Africa is not economically independent. What is the value of political independence if there is no economic independence? Which recent year has African countries been able fully fund their budgets? We continue to look up to other countries, which we are not ashamed to call donors partners for the revenue shortfalls in our budgets. It is often said, the one who asks for help should be ready for terms and conditions of the helping hand. That is why Africa's donors continue to dictate for Africa what to do even if it will be to the continent's disadvantage. Now, African countries are running away from the "visible" strings attached loans from the International Monetary Fund (IMF), the World Bank, and other Western countries. But where do these runs end up—at the loans with "invisible" strings?

China has become Africa's new friend. At least in the minds of African leaders. China has chosen not to make any irritating demands popularly referred to as "conditionalities" on African countries and governments regarding her assistance to African countries. China is ready to give billions of loans to the countries of Africa provided they can guarantee payment with the rich natural resources or a strategic asset.

In the time of Ghana's president John Evans Atta Mills, China promised to give Ghana up to $15 billion of loans and grants when the former president visited the country in 2010.[E1] Currently, the government of Ghana is considering floating a $50 billion century bond with China.[E2] In July 2018, the parliament of Ghana approved an international agreement with China that will give Ghana $2 billion in exchange for Ghana's bauxite amid concerns from some minority members of parliament.[E3]

It is clear that the path Africa has chosen to tread on is a path of dependency on other countries for its survival – not for progress. Africa has chosen to be dependent on other countries because it cannot raise enough money to fill the holes in its budget. Africa needs to raise money to build the economy's infrastructure, which is lacking. Interestingly, I do not see any long-term plan from African countries to wean themselves out of the way of economic dependency shortly as it continues to tread on the paths that make them more dependent.

The issue facing Africa then is, whom should Africa depend on in her quest for development? Should Africa depend on the Western countries or her new friends in the East? Over four decades, Africa depended on the "visible strings attached" funds of the Western countries to meets her financial needs. Under such funding, the West called on the richly resourced but poor African countries who are unable to manage their resources effectively to be financially disciplined. Or they will be required to raise more revenue or respect human rights or adhere to good governance or whatever they see as wrong with the country before they decide to offer such assistance or bailout the country may need. The rules governing the new "invisible strings attached" funds of recent years offered by relatively new economic powers of the East are different. The countries of Africa are "sovereign," and they elect their leaders. The leaders can do whatever they want in their own countries because they are accountable only to their people. If Africa needs money, it will be given to them. The primary concern is that Africa has to pay back the money when it is due. So far as Africa has natural resources or other assets that can pay for the loan, they can take whatever loan they need for their sovereign states.

Comparing the "old" and "new" order for funding Africa's infrastructure deficit, the "old order" was controlling how African countries govern their countries. To the West, because African governments have proven that they are reckless in expenditure and lazy in raising revenue or converting its natural resources to financial resources to provide for the needs of her citizens, Africans must be directed in one way or the other to ensure economic prudence. With the "new order" from the East, Africa has no interference with how it governs itself. It can be as reckless as it wishes, but should Africa become so reckless to the point that it cannot pay for the monies it borrowed, then Africa has to sell or give away her natural resources to offset the loans it took. In Ghana, the old term "batter" is the new term that is being used to describe such deals. The question that needs to be answered is, which of these is the best option for Africa? This question should not be left to African leaders alone to answer; the entire population of Africa has to answer the question.

Africa will continue to be dependent on other countries and make mockery of its political independence if it is unable to raise money out of its resources to provide for every basic need of its citizens. Many Africans wonder what African leaders do at the African Union meetings. There are many who share the view that when the Organization of African Unity (OAU) achieved its objective of getting Africa free from colonial rule in 1994 with South Africa being the last African country, OAU lost its relevance and needed to fold up. Proponents of this school of thought believe that, though the OAU had the long-term economic independence of Africa in mind, the purpose, structure, and orientation of the organization were more toward attaining political freedom for the continent. Therefore, simply changing the OAU to AU was to mimic the success story of the European Union, and that explains why the AU has not been able to do much when it comes to the economic integration of the continent and its pursuit of economic independence.

At a recent forum on China-Africa Cooperation (FOCAC), China's president Xi Jinping invited the entire AU virtually except for Swaziland, which is Taiwan's only African ally, and promised to give Africa over $60 billion in loans and grants.[E4] This is certainly welcome news for African leaders who need money so badly but have the money right under the soles of their feet. It is a piece of good news because when Ghana needed something less than a billion dollars to support its economy from the IMF, it was challenging to get that money, which was released in tranches over three years. It's not a wonder that the president of South Africa defended China at the 2018 FOCAC that they "refute the view that new colonialism is taking hold in Africa as our detractors would have us believe."[E5]

The bottom line is that it is plain and clear that the conditions that are attached to the "East funds" might not inure to the interest of the African just like the "West funds" that Africa is now trying to avoid. Many African countries had to go HIPC because they were not able to pay for their indebtedness to the West. In an asset-backed loan, the country might have to surrender an asset on default.

Zambia and Kenya are some of the countries that have been in the news in recent times because of high indebtedness. Zambia's external debt is $9.4billion as of June 2018 with China being a major lender to the nation.[E6] Indeed, it was reported by *Africa Confidential* publication on September 14, 2018, that Zambia's indebtedness to China is a point where there are talks between the two countries about a possible takeover of Zambia's national electricity company, ZESCO.[E6] The essence of raising this point is that, just as businesses borrow money for investments that empowers them to pay back the loan, Africa's borrowing should be able to lead to economic independence in the future. Africa has been on the borrowing path before, which led it to the HIPC initiative where billions of her debts were written off. Africa should therefore be wary of excessive borrowing to avoid another form of HIPC because this time, the debts may not be written off. Africa may have to give her resources in exchange of its debt. Africans have to ask themselves many questions in her quest for development. Africa needs not swallow anything that is thrown out there. Asking questions might bring out some relevant answers.

THE ABANDONED DIFFICULT PATHS

Can Africa be economically independent? More importantly, is there a better option for African economies? I believe Africa can emulate the "Asian Miracle" just that Africa has refused to follow the long and arduous path. Instead, Africa has chosen the short and easy way, which is, unfortunately, taking Africa nowhere. Africa is fortunate to have some countries to look up to and learn, but it just will not learn. Korea and Taiwan, for example, focused on building human capital and systems to take them from third world countries to first world countries. They had little natural resources, but they made it. Can Africa have an excuse or any reason to remain a third world country? I will be glad to know. Africa doesn't want to make any sacrifices and yet wants to have it all. Every nation that has been able to attain fast and sustained economic growth made some sacrifices. Africa's new friend China sacrificed larger families and kept the one-child policy for the good of their nation. Singaporeans used to sacrifice and continue to forgo their salaries as at least 30 percent of their incomes is set aside by their government, which is managed for a better future in retirement.

We are so eager to borrow than to pursue policies that will give us steady and sustained growth. African leaders have abandoned the brains they used the taxpayers' money to pay for at the various universities and higher institutions of learning and are doing what they think is right, while such institutions and human resources are the keys toward finding a lasting solution to the present and future challenges. When will Africa learn? The leaders in Africa always want the easy way out, which is to borrow. For it is easy to borrow money to spend than to work for money. When you work for money, you need some discipline as the money will trickle in, in amounts that have to be managed effectively to perform the tasks required. But when you borrow, you can get all the money you need within the shortest possible time to do what you want to do. The question is, have we assessed the long-term cost of the loans to our children and the next generation? I don't want to even think about the interest that will accrue, but I am looking at Africa's control of its natural resources. If Africa continues on this trajectory, what will become of us as citizens when the limited resources are depleted? At least for now, we are the envy of the rest of the world because of the natural resources we have. The only thing we now have to contribute to better the world is our rich natural resources, and we are losing it at an incredible rate.

It is time Africa restrategize to meet her current and future needs because dependency cannot solve Africa's problems. Borrowing in itself cannot solve Africa's problem. Only strategic thinking that requires sacrifices today to build on our strengths and progressive cooperation from other countries will. Africa has gone into the periods of borrowing before, and it didn't help that much. Yes, we need the infrastructure to open up the economy, and all the right reasons for borrowing are good. But is borrowing the only way out? Is borrowing the only tool used by other countries to establish their infrastructure? If it is the only way out, then let's go ahead; but if it is not and there is a better option, then the next generation of Africans is looking up to this generation to make that decision in their interest.

Africa borrowed heavily and until the debts became unsustainable. We were at the mercy of our creditors. The creditors, the World Bank, and IMF had to show compassion and allowed countries to subscribe to HIPC so that some of the debts can be forgiven. Many countries including Ghana accepted the deal because there was no choice, and the economies were suffocating from servicing debts and interest payments. Other countries such as Japan, Germany, etc., made some bilateral arrangements regarding Africa's indebtedness to their respective countries, and that gave African countries some more relief. Is Africa going to repeat this same cycle? We don't need a soothsayer to tell us that the policies in place now will not be able to pay our debts. The conditions attached to those debts should, therefore, be of concern to the leaders because, like my country, Ghana, I can only imagine what will happen to us when we deplete our natural resources because even with these resources, many of our residents live in poverty and hardships. How much more when we have nothing?

I became more worried when I read an article in the *African Exponent* written by Sebastiane Ebatamehi on September 29, 2018, with the title; "Zambia Deported Renowned Pan-African, Prof. PLO Lumumba, for Speaking Up Against China."[E7] The article had it that the professor was to deliver a lecture at the Eden University on the topic, "Africa in the Age of China's Influence and Global Geo-Dynamics," and was denied empty on arrival on the grounds—that the learned professor was a security threat to the nation of Zambia.

Though I am not privy to the kind of security threat, but as a fellow in the academia, I know what we do. Academicians are passionate because we seek and find information and make sure that those taking decisions regarding our welfare base their decisions on it. Academicians also make sure that everyone has access to that information because that is what we seek state funding to do. At times, it is worrisome to go through the hazards of securing funding to seek for information and not disseminate it, while people are committing mistakes because they are ignorant of such information.

As I said earlier, I am not privy to the information, but whatever it is, we can also guess what it may be. The professor has been to Ghana and spoken at events and was not a security threat to Ghana. The professor has expressed his concern over the "Chinalisation of Africa." His concern, as I have heard him, is not hatred for any country but rather the need for Africa to protect Africans as it relates to other countries. If Pan-Africanists are now security threats to Africans, then there must be something wrong with Africa. Africa, which used to be a safe haven for Pan-Africanists like W. E. B. DuBois and other Africans in the Diaspora, is now a no-go area for Africans that are passionate about Africa. "Victims," song of Lucky Dube, came to mind when the news broke about his assassination in South Africa.

"Bob Marley said; How long shall they kill our Prophets while we stand aside and look. But little did he know that eventually, the enemy will stand aside and look while we slash and kill our own brothers."[E8]

I could hear the words again as I read the article. Africa is on a path—a path to nowhere; but I guess it may be a path back into the dark dungeons of slavery, but this time with our brothers and sisters holding the ends of the chains that bound us. In these dungeons, there will be no ships coming or waiting on the sea. If there were, there will be hope—hope that we will see the light of day again. Hope that there will be land to toil on—a land where we can fight for recognition and equal rights for our voices to be heard. These are the dungeons of our land—the dungeons whose architect, builders, gatekeepers, and guards are people of our color and blood.

E1   Connors, W. (2010). China Extends Africa Push With Loans, Deal in Ghana. Accessed December 12, 2018 through https://www.wsj.com/articles/SB10001424052748703384204575509630629800258

E2   Nyavor, G. (2018). Ghana considering floating rare $50bn Century Bond. Myjoyonline.com. Accessed December 15, 2018 through https://www.myjoyonline.com/business/2018/september-3rd/ghana-considering-floating-rare-50bn-century-bond.php

E3   Ghanaweb (2018). Cash for Bauxite: Parliament okays $2bn Ghana-China barter trade. Accessed December 12, 2018 through https://www.ghanaweb.com/GhanaHomePage/NewsArchive/Cash-for-Bauxite-Parliament-okays-2bn-Ghana-China-barter-trade-673389

E4   Aljezeera (2018). China's Xi offers $60bn in financial support to Africa. Accessed December 12, 2018 through https://www.aljazeera.com/news/2018/09/china-xi-offers-60bn-financial-support-africa-180903100000809.html

E5   AllAfrica (2018). Africa: South African President Rejects Claims of Chinese "Colonialism" Accessed December 15, 2018 through https://allafrica.com/stories/201809040284.html

E6   Later, V., & Mususa, P. (2018). Is China really to blame for Zambia's debt problems? Accessed December 11, 2018 through https://www.aljazeera.com/indepth/opinion/china-blame-zambia-debt-problems-181009140625090.html

E7   Ebatamehi, S. (2018). Zambia Deported Renowned Pan-African, Prof. PLO Lumumba, for Speaking Up Against China. Accessed December 12, 2018 through https://www.africanexponent.com/post/9177-kenyas-professor-patrick-lumumba-denied-entry-into-zambia-deported-back-to-kenya

E8   AZ Lyrics (2018). Lucky Dube Lyrics. Accessed December 16, 2018 through https://www.azlyrics.com/lyrics/luckydube/victims.html

CHAPTER SIX

Education: The Path To Aliens Citizenry

Knowledge is power. Information is liberating. Education is the premise of progress, in every society, in every family.
—Kofi Annan (former UN secretary general)

The statement above is the words of Kofi Annan, a true son of Ghana who rose up to become the boss of the United Nations. If his words are true, then education was to be the foundation of progressive Ghana and other independent Africa states when they became independent. The pivotal role of education is echoed by Christine Gregoire, who asserted that "education is the foundation upon which we build our future."

The British colonial government of the Gold Coast initiated the educational system in Ghana and set up the first schools in the country. However, the schools at that time were meant to pursue an agenda. The agenda was to meet the educational needs of the colonial government. It was expected to facilitate the then government's efforts at controlling the African people and taking away the precious minerals of the land. Schools were intended to train few Africans to be interpreters and to think and understand the white man so that they can be used against their fellow blacks. The few Africans who had access to the schools were to be trained and pitched against their fellow blacks so that they can convince their fellow blacks that the white man has the interest of the black man at heart. In that way, the white man will have the support and cooperation over the black man over which they (whites) rule.

It must be understood that purpose or aim determines everything. Everything the creator or man who is created in the image of God does or creates is for a purpose. It is the purpose that drives the action or the design if it is an object. Airplanes are designed to have wings based on the purpose that it is supposed to fly. Cars have wheels because they are designed to move on the ground from one place to the other. The chairs are designed to give comfort to the one who will sit on it. Nobody sits on a chair with the intent of moving from one place to another. The one who does that will remain stuck to where he is.

Once a purpose for which something is to be made is set, it influences the design of that thing so that the design will perfectly match the environment or the conditions under which it will operate and how it can overcome the challenges confronting the achievement of that purpose. For instance, since humankind had the purpose to fly, then something had to be designed such that it can overcome the challenges of gravity, wind resistance, turbulence, extreme colder conditions associated with height, the power to sustain it in the air, controls that can send the airplane up and down as needed, etc. The whole design of the aircraft is meant to serve that purpose. Military aircraft, choppers, etc., also do fly, but because they serve a different purpose, they have also been designed differently to serve that purpose.

The reason for making the comments above is that the educational systems in almost all African countries were set up by colonial governments, and they had a purpose. The purpose was not meant to develop the young African to be creative and think for themselves, to acquire and develop skills and prepare them to overcome the challenges of life, and to pursue their passions and have that confidence to fulfill that purpose. The purpose of education as indicated earlier was to serve the educational needs of the colonial government. Thus, the educational system was designed to provide an education to the black man at the time so that the black man becomes a tool in the hands of the colonial government that can be manipulated to achieve the broader goal of the white man.

After colonial rule ended in the Gold Coast (Ghana), there was, therefore, the need to retool Ghana's education to serve the purposes of the freed black man. The kind of training that will make the black man capable of managing his affairs. The kind of education that trains the black man to trade with the white man and make profits from his trade. The educated black man who can negotiate and get a fair share of the world's resources. The educated black man who will ask the reason why the white man was bent on taking the resources of the black man's land. The educated black man who will not betray his fellow black man for favors from anyone. The educated black man who will understand the behavior of his fellow black man and has an interest of the fellow black man at heart. An educated black man who will no more seek to serve the white man but his fellow black man. A black man who will not see a fellow black man of a different tribe as a foreigner. An educated black man who will seek knowledge about his environment, not the environment of foreigners.

Unfortunately, the black man continued with the educational system bestowed to him by the colonial government with little educational reforms that did not touch on the foundation laid during the colonial era. As a result, today, African schoolchildren continue to use books with Jack and Jill as characters in schools. African schoolchildren continue to sing the choruses of "in the bleak midwinter" and read about snow when they may never see snow in their lives. These kinds of education were not designed for the liberation and progress of the black man. It was designed for something else, and unfortunately, Africa has failed to redesign education to serve the African. It will be difficult for the children of Africa to successively deal with the challenges of life and realize her objectives because the system that was designed wasn't designed to deal with the issues of the liberated African, and it has not been redesigned. In essence, the African child is on a bicycle and pedaling as fast as he can with the hope that the bicycle will fly and glide in the skies one day. It is an aspiration that will never materialize until the bicycle is redesigned to make it fly. For education to take the African to where we want to be as a country or continent, it needs to be redesigned to fit the needs of the country or continent.

EDUCATIONAL TENSIONS AND INHIBITIONS

Education in Ghana, the star of Africa, started on the wrong note. The colonial government set up the educational system for the purposes of training the African elite who were to facilitate the exploitation of their brothers and sisters. They were to be interpreters for the colonial government. The educated Africans were to be offered certain privileges that were to make him a friend of the colonial master and an indirect enemy to the cause of the fellow African. In essence, the educated black man was to be the one who could speak the language of the white man. He was to be the one who will think and act like the white man and be able to understand and support the cause of the white man. He was to be the one who can facilitate the business of the white man, and the business at that time was mainly to export the rich natural and human resources of the African people.

According to historians, the primary purpose for teaching the youth of the local people was to provide them with the skills needed for employment in the booming European commercial enterprises along the coast of the Gold Coast. Attention concerning education was not given to the challenges, issues, and concerns of the black man. Research, resources, and efforts were not invested in promoting the cause of the African people. The educated African therefore had no essential skill that made him ready or prepared to take on the present and future challenges of the continent. Instead, he had all the skills and training to serve the needs and meet the aspirations of the colonial government. The educational system, as pointed out by the Phelps-Stokes Commission on Education in Africa report, had no intentions of educating the Africans but just a few of them.[F1] The masses were to be left ignorant, and the few educated were to serve a special purpose. The educational system created a class society where the few Africans who had the opportunity to be educated became the elite group who were to be lords over those who were not.

When Ghana became independent, it had its present and future challenges. Unfortunately, according to the status quo, educational curriculum and structure were maintained. This was the beginning of the tension and inhibitions that will be the bane of education in Ghana and other African countries.

THE COLONIAL STYLE OF EDUCATION

It was imperial instructions from Europe that encouraged governors to start giving instructions to the local people of the then Gold Coast in the eighteenth century. Teaching of the selected few took place at the castles with Elmina, Cape Coast, and Christiansborg castles in Accra being the center of the Western-style education. Those who received instructions were carefully selected, and they were the offspring of the wealthy African merchants who were trading with the colonial government and providing for the needs of the government. The children and relatives of other local chiefs who were transacting business with the white man also qualified for instructions. So were the "mulattos," children of the native African women who were impregnated by the white man.

The Europeans increased the number of schools in the nineteenth century as more missionaries got to the Gold Coast and started their schools. The missionary schools at that time also had the primary purpose of spreading the Gospel of Jesus Christ and needed specific skills among the local people to achieve that objective. However, their attempts at extending the Western-style education from the coast toward the interior parts of the country, which were a territory of the Asante Kingdom, were met with fierce resistance. The Asante Kingdom's opposition was based on their belief that the Western-style education will not only be detrimental to the course of the black man but also impact negatively on local values.

There were conscious efforts to limit practical training in agriculture, crafts, and talent development among the local people. This was echoed in the Phelps-Stokes Commission on Education in Africa report issued in 1922, which pointed out that the system of education in the African colonies was deficient in meeting the economic and social needs of the continent and was also not improving the conditions of the majority of the African people.[F1]

Though there were pressures from Africans in the Diaspora and recommendations suggested to make education more relevant to the course of the people of the Gold Coast through the inclusion of industrial and agricultural training, little reforms were undertaken by the colonial government. According to Foster, "Western education (had become) the most visible and tangible manifestation of European power, hence access to that power demanded entry to the type of education provided in the metropole itself."[F2] Sir Hugh Clifford, in 1918, set some educational targets that intended to broaden at least primary education to every child in the Gold Coast.[F3] The targets included the setting up of teacher training institutions and a college to provide secondary and postsecondary education. Regrettably, the educational development targets remained targets and never saw the light of day as education continued to be a privilege of the few.

Emphasis continued to be on the academic education, which trained the African to occupy the high-paying white-collar jobs of the time and agriculture, and industrial training was relegated to the background. And looking at the way it was at that time, if the reward for academic education was that high because those jobs were paying well, obviously, people who had access to education at that time will all prefer to pursue academic education with the goal of becoming employed in the lucrative businesses along the coast of the Gold Coast.

The repercussions of emphasis on academic education have been with the Ghanaian and the African till today. As of now, our universities continue to churn out people who are looking for white-collar jobs and not the people with the skills to do business by providing solutions to the everyday challenges the people of Africa face. Under the colonial system, only a few had access to education, making the supply of human resources to feed the vacancies in the white-collar jobs small. As the country gained independence and extended the same academic education to the masses, the supply of human resources to the white-collar jobs became much greater than the vacancies available. This is another reason why we have such high levels of unemployment among the youth of Ghana and by extension Africa today. Thorns suck up the nutrients in the fertile lands of Africa, while graduates from agricultural universities roam the cities of Ghana and other African cities seeking white-color jobs that are nonexistent.

EDUCATION IN POSTCOLONIAL GHANA

Nkrumah's Era to the Rawlings Era (1957–1981)

The first Ghanaian government administered by Dr. Kwame Nkrumah initiated the first educational reforms through Act 87 of 1967. Like the targets set by Sir Hugh Clifford, it sought to achieve free universal primary education. For the first time, real efforts were made through the act to ensure that Ghanaian children from age six have no excuse for not being in school because fees were not charged, and educational infrastructure was expanded to accommodate children of school-going age. The educational system started with primary education for six years and a four-year secondary education, after which those who would want to pursue further education had to take a sixth form program for two years that can lead to a three-year university program.

The new educational system in the '60s continued on the lines of the precolonial era by producing skills needed for employment in white-collar jobs as these jobs continued to be lucrative. The system was mainly academic with little focus on vocational and other skills. Others criticized the length of the educational period for being too long and wasting a lot of youthful and productive years of the youth of the country.

Nkrumah was overthrown in the 1966 coup by the National Liberation Council (NLC) led by three military leaders, Lt. Gen. Akwasi Amankwaa Afrifa, Lt. Gen. Joseph A. Ankrah, and Lt. Gen. Emmanuel Kwasi Kotoka. There were no significant educational reforms under the National Liberation Council (NLC) governments rule from 1966 to 1969. Ghana returned briefly to constitutional rule after the elections of 1969 and the Progress Party led by Dr. Kofi Abrefa Busia. The PP's government, like the previous administration, did not introduce any significant changes in the educational system. Another military leader, Gen. Kutu Acheampong, sent Ghana to back to military rule in 1972 when he toppled Dr. Busia's government.

Acheampong's National Redemption Council government initiated educational reforms in 1974 by establishing the Junior Secondary School (JSS) as a way of moving from the academic-oriented educational system. The JSS, which was introduced on an experimental basis, brought practicality into the educational system by helping students to acquire occupational skills. The JSS system remained experimental and was never fully implemented because of many factors including rapid economic decline and bureaucracy. Successive governments after Acheampong from 1975 to 1981, including the Supreme Military Council (SMC), the Armed Forces Revolutionary Council (AFRC), and PNC governments, did not make any significant changes in the educational system. The period saw a decline in the quality of education as educational infrastructure deteriorated, and dropout rates reached alarming levels with conditions going from bad to worst.

The Rawling's Era (1981–1992)

Flt. Lt. Jerry John Rawlings, who overthrew the SMC military government on June 4, 1979, and instituted the democratic government of the PNC in September the same year, returned the country to military rule on December 31, 1981. Rawlings inherited an educational system that was in a mess. The state of education in Ghana by 1983 was that of a crisis. Funding to the sector was limited, enrollment was low, and basic educational supplies were lacking.

Rawlings came out with another educational policy for Ghana, which was known as the Education Reform Program (ERP) in 1987, with the support of the World Bank and other donors. The reform primarily shortened the number of pre-university years from seventeen to twelve years and emphasized on vocational education. The JSS system was reintroduced, this time on a national scale. Primary education was extended from six years to six years of primary and three years of junior secondary education. The ERP designed basic education to ensure that children at the basic level become creative and acquire skills and attitudes to overcome challenges in their environment.

After reviewing the success of the reforms, the government moved a step further to introduce the Basic Education Sector Improvement Program (BESIP), popularly called the Free Compulsory, Universal, Basic Education Program (FCUBE), as a way of increasing access and quality of education at the primary level. From JSS, students progress to a three-year secondary education, after which they have the option of four-year university education, a three-year teacher training education, a three-year nursing or basic health education, or a three-year polytechnic education.

The Fourth Republic Era (1992–present)

The Fourth Republic, or Ghana's return to constitutional rule, began in 1992, and the structure of education has not seen any major reforms since the ERP of 1987. The introduction of the Ghana Educational Trust Fund (GETFund) has helped to improve infrastructure and access to education. Private individuals and religious bodies have also contributed to improving access to higher levels of education with the setting up of private universities and research institutions.

Vision 2020, which is Ghana's long-term program toward achieving a middle-income status nation, has education as a major player in that drive. The vision aims at making the entire Ghanaian population more productive, which requires that all Ghanaians acquire basic education. The policy further aims at improving the levels of science and technology education among its population.[F4]

I have discussed Ghana's educational experience since independence as an example. Almost all African countries have tread on educational paths similar to that of Ghana. Governments after governments have not prioritized education and have failed to execute reforms that will address the educational fundamentals. Conflicts and wars have destroyed educational infrastructure, and governments have to resort to borrowing to fix the deteriorating infrastructure.

THE WINDING CURVES ON THE EDUCATIONAL PATH

From the brief review of the educational plans, programs, and policies over the years, it is clear that Ghana's educational system has had challenges, and successive governments have not been able to address those challenges. Indeed, the challenges have been compounded, requiring more efforts to get back on a strong footing.

Dr. Nkrumah, though, made the right decisions by eliminating the affordability barrier and did not make conscious efforts to make the contents relevant to the African. Drawing his inspiration from the then Union of Soviet Socialist Republic (USSR), it wouldn't have been too hard to for him to balance the purely academic kind of education with other skills and technical or vocational education. Education in Ghana from independence has centered on training for employment, and that trend has continued till today. The educated Ghanaian is not an industrious one. The educated Ghanaian is not trained to employ others; he is trained to be employed. To absorb the heat of unemployment, African governments have resorted to creating numerous government agencies to employ the teeming graduates who do repetitive and unproductive tasks. The wage bill is sapping higher percentage of government revenues. The South African Government News Agency in 2012 reported that government wage bill took 42 percent of government revenue.[F5] The wage bill took 45 percent of Ghana's tax revenue in 2017.[F6] Data from the IMF indicates that on the average, the wage bill takes 32 percent of government revenues in sub-Saharan Africa.[F7]

Advisors of many African economies have asked for reduction in governments' wage bill. The bigger question is, who is going to employ the educated Ghanaian when the "irrelevant" government agencies are scrapped? There has been the oversupply of people who are ready to be employed and undersupply of the people who will employ the educated ones. Ghana and other African countries that followed on such paths needed no prophets to predict that African countries were on their way to unemployment, and this time graduate or professional unemployment.

In the case of Ghana, successive governments did not rectify the colonially induced problem in Ghana's education and continued with the status quo until the 1974 reforms by the I. K. Acheampong's government. Though Acheampong saw the actions of the West and other institutions of the West as a subtle drawback to Ghana's path to economic independence, the approach of cutting ties with countries and institutions was not the best, as it compounded the problem in the educational sector.

Eventually, Acheampong's policy toward the West backfired in such a way that his attempts to correct the imbalance in the system did not materialize mainly because of lack of funding to the sector as a result of the bad state of the economy resulting from the breaking of ties with other important global players. Breaking ties is not always the best option. There is always room for renegotiation and reengagement. If circumstances are not favorable to a party in a deal after it has been entered into, there are always opportunities to reengage on more favorable terms than severing ties. In a global economy, countries, institutions, and organizations, no matter how small they may seem, have the potential to affect nations by their actions. It is important that rulers of nations understand how to get the best out of deals than to walk away from them.

Rawling's educational reform, ERP of 1987, undoubtedly made an impact but not the best of impacts. The implementation of the program started well with good intentions, but it lacked funding and sustainability mechanisms. As a young boy growing in the village of Akomadan, I saw Junior Secondary Schools fully equipped with carpentry and other vocational skills tools, which helped pupils to construct tables and chairs and other basic things made of wood. Though I was in a private school at the time, and private schools used to vacate earlier than the public counterparts, I always enjoyed coming from Kumasi to see my colleagues in the public school next to my house exhibiting their creativity, which I envied. Currently, none of the tools or practical workshops exist in our JSS schools in my town and the entire district.

Thus, basic issues that were standing before independence in Africa's educational system continue to exist.

- Focus is mainly on academic education with the other forms of education given limited attention and resources.
- There is difficulty or lack of opportunity for those who choose the other forms of education like technical and vocational education to pursue higher knowledge of degrees.
- Lack of sustainable funding to support educational programs has been a challenge. The educational system in Ghana has continued to rely on funding sources it has no control over. Cutting those funding sources, which can come at any time because of unfavorable conditions in the donor country, will mean doom for our educational system. I remember the challenge the Ghana School Feeding Program went through when funding from the donors ceased. Any educational policy

that is not supported by sustainable funding will not achieve its intended purpose.
- It takes too long a time for the country to review educational policies and curricula. The world is moving at a faster rate, and the dynamics keep changing. Our educational system must be in tune with the changing dynamics so that it can produce the human resource that is skilled enough to take up the current and future challenges with a greater percentage of success. Related to this is the curriculum review. When the external environment is well monitored as the trend of skills needed in the future, it will automatically inform the review of the curriculum to meet that skill requirement.
- Lack of enforcement behind educational policies. The country has never fully enforced its educational policy. Primary education, if free, and governments have provided free textbooks, free school uniforms, and even free meals in our primary schools, but children of school-going age continue to roam on the streets of Ghana aimlessly at the watch of policy funders, the taxpayers.
- The lack of settlement policy is another bane in Ghana's effort at improving education. Ghana is always overstretching its limited education budget with unplanned settlements. Unlike other countries, Ghana does not have any clear-cut settlement policy. A family will create a farmhouse for his farming activity, and very soon, that develops to about ten houses, and they will put up a makeshift structure, gather the children, and begin a school, which it will apply to the district education office to absorb. For political purposes, such schools are absorbed with nothing planned for educational materials, teachers, etc. A clear settlement policy will help the country avoid such expenditures and also ensure that the children who get into the schools have all they need for effective teaching and learning.

It is evident from the discussions above that both precolonial Ghana and postcolonial Ghana has had its educational policies strongly influenced by external forces—forces whose interests are very different from the desires and aspirations of the African people. Even today, the African continent still has its educational policies dictated by the countries and institutions outside Africa. Ghana, like many other African countries, signs onto international agreements such as the Millennium Development Goals and graft in policies of education whose adaptation and future repercussions on African values, culture, and progress may not have not been properly scrutinized.

A major reason behind Africa's inability to take bold policies devoid of the dictates of international organizations, countries, and institutions is the fact that it is unable to support the home-grown educational policies with the needed sustainable sources of funding. This is where Africa and the friends of Africa have to come together to find sustainable ways of mobilizing funding to support long-term educational plans that will educate the next generation of Africans to realize the African dream. Education is valuable, and it is expensive, so countries need to take time to prepare a sustainable source of funding to educate her citizens. Until the people of Africa can get its educational funding right and devoid external control as a country, regional body, or continent, the African dream will remain elusive. Every educational policy has to be backed with sustainable funding not from outside the continent but from within it. Until then, education in Africa will not center on addressing the core needs of the African people, and Africans will not be able to do much about the situation in Africa, but, rather, foreigners will be in charge of what happens on the African continent.

EDUCATING AFRICA WITH A FOREIGN LANGUAGE

Unlike other nations, the boundaries of many African countries including Ghana were set in place in Europe.[F8] As a result, various tribes that happened to be within a particular geographic area at a point in time became citizens of one country. The Asantes of Ghana, for example, are people of one language and culture. However, the demarcation of the borders of present-day Ghana cut off some people of the Asante tribe from Ghana. The Asante traditional territories lie beyond the borders of present-day Ghana extending into modern-day Ivory Coast and Burkina Faso. Those Asantes outside the borders of Ghana still pay homage to the king of the Asante Kingdom, and during major festivals, the Asantes living in other countries come "home" to serve their king.

The demarcation of the boundaries of African countries at the Berlin Conference resulted in countries that have many tribes and languages coming together to form one African country. African countries were not derived from people of one country and culture coming together to pursue one agenda. Rather, people who speak different languages, have different beliefs, and of different cultures were made to coexist or come together to form one country and controlled to pursue the agenda of another country. That is how the countries of Africa was formed. This is very different from what happened elsewhere on other continents. The forceful coexistence is a major reason behind the many tribal wars that have continued to disturb the peace of many African countries even until today. The genocide in Rwanda was largely attributed to differences between the two major tribes that were made to coexist to form the country.[F9]

The boundaries of Ghana knitted together over seventy ethnic groups to form the present-day Ghana. As a result, there are over seventy spoken languages in Ghana, though only nine are government sponsored, which are spoken, written, and considered major languages. This was certainly going to be a challenge for the country at independence because of the difficulty of having a national language that will bind the people in the country together. The situation compelled the founders of the nation to adopt English as the official language and a medium of instruction in schools when the Europeans left the shores of the country.

The other newly Independent countries of Africa faced similar challenges concerning which tribal language to adopt as official languages. In the end, the newly independent African countries adopted the language of the previous colonial government, leading to different countries in Africa speaking languages such as English, French, Portuguese, Spanish, German, and Dutch. In the case of Ghana, Ghana adopted English as the official language, while Ghana's neighbors, Cote D'Ivoire on the west, Togo on the east, and Burkina Faso on the north, all spoke French. At the south of the country is the Gulf of Guinea of the Atlantic Ocean. In effect, Ghana's ability to trade or integrate with its neighbors was given a serious blow at birth. To this day, Ghana's inability to effectively trade and undertake other economic activities with her neighbors has been attributed to the inability of her citizens to speak or understand the language of her neighbors.

Fundamentally, it was to be expected that the French and English languages, which were the languages of most of the colonized West African countries, should have been introduced in Ghana's curriculum at the primary level to ensure that the next generation of Africans was able to communicate among themselves within and outside their own countries. This was not carried out with the seriousness it deserved, and various reforms in Ghana's educational system failed to prioritize it.

Casting our minds back to the beginning of the European Economic Community (EEC), which metamorphosed into the European Union, we get an idea of how Africa's lackadaisical attitude toward the issue of languages impeded Africa's economic integration among itself. Few days after Ghana had her independence, six countries in Europe came together to sign the Treaty of Rome, creating the European Economic Community (EEC), or common market, on March 25, 1957. The countries spoke different languages including German, Italian, French, and Dutch, and they foresaw that the economic integration would not be successful without breaking the existing language barrier. Europe set the policy goal at making sure that every European speaks two other languages in addition to their mother tongue. Each country in Europe treated the policy with the seriousness it deserves, introducing two other European languages to the European child at an early age.[F10] Today, I can move across Europe, from Portugal to Poland, because I speak English. It is interesting the response you receive when you enter a shop in Spain or Germany or Switzerland, and you ask, do you speak English? The response, as if they have been taught by the same teacher, is "A little bit." When you begin the conversation, you will find out that they can express themselves very well in the language they know "a little bit" about.

Africa was unfortunate to have colonialism, which not only established present African states but also bequeathed official languages to African countries. Once African countries embraced the languages of former colonial governments as official languages, there was the need to make sure that those official languages become the medium of bringing African countries together. Africa needed to focus on making sure that her future generations were introduced at an early stage to at least two of languages such as French, English, Portuguese, Arabic, German, African, Swahili, and Spanish. Economic integration in Africa wouldn't have been the same as we see it today. Personally, I lost two major opportunities to head regional offices of major international organizations because of my level of proficiency in the French language.

There are many Africans and Pan-Africanist who are proposing that Africa can only move forward if it can go back to its own languages and educate her children in their mother tongue, though I have no objections to the ideas and believe that a common language is essential. I have experienced learning in a mother tongue and learning using English as a second language. Personally, I would have found it to be more convenient if I had my instructions in science, mathematics, etc., in my mother tongue. But it is not easy for Africa to return to its mother tongue. Using Ghana as an example, using Akan as the official language of the country will mean that Ewes, Dagombas, Gas, etc., will have to learn Akan as a second language. In that case, we will still have a lot of Ghanaians learning subjects using a second language as a medium of instruction. Using English, which has the additional advantage of being an international language, will have other additional benefits of being able to communicate extensively in a global economy. Africa has to bear the difficulty of learning with a foreign language as a medium of instruction, but it has to push the next generation to speak more of its official languages for deeper integration of the continent.

F1 Berman, E. H. (1971). American influence on African education: the role of the Phelps-Stokes Fund's Education Commissions. *Comparative Education Review*, 15(2), 132-145. http://www.jstor.org/stable/1186725.

F2 Forster, P. (1965). Education and Social Change in Ghana. Routledge & Kegan Paul. London

F3 Prices Ghana (2019). History Of Education in Ghana. Accessed February 7, 2019 through https://pricesghana.com/history-of-education-in-ghana/

F4 Republic of Ghana (1995). Ghana Vision 2020 (The First Step 1996-2000). Presidential Report on Coordinated Programme of Economic and Social Development Policies (Policies for the Preparation of 1996-2000 Development Plan).

F5 IMF (2012). South Africa: 2012 Article IV Consultation. IMF Country Report No. 12/247. Accessed January 2, 2019 though https://www.imf.org/external/pubs/ft/scr/2012/cr12247.pdf

F6 Ghanaweb (2018). Wage bill hits GHC14bn, takes 45% of tax revenue in 2017. Accessed December 11, 2018 through https://www.ghanaweb.com/GhanaHomePage/business/Wage-bill-hits-GHC14bn-takes-45-of-tax-revenue-in-2017-671387

F7 Clements, B., Gupta, S., Karpowicz, I. & Tareq, S. (2010) Evaluating Government Employment and Compensation. Washington DC, International Monetary Fund, Fiscal Affairs Department.

F8 Herbst, J. (1989). The creation and matintenance of national boundaries in Africa. International Organization, 43(4), 673-692. doi:10.1017/S0020818300034482

F9 BBC (2011). Rwanda: How the genocide happened. Accessed December 11, 2018 through https://www.bbc.com/news/world-africa-13431486

F10 European Commission (2017). About multilingualism policy. Accessed December 16, 2018 through https://ec.europa.eu/education/policies/multilingualism/about-multilingualism-policy_en

CHAPTER SEVEN

Corruption: The Path Of Individualism And Division

This land of ours cannot be a good place for any of us to live in unless we make it a good place for all of us to live in.
—Richard Nixon

Corruption is a difficult problem that every country has to face. Many are of the view that corruption is unique to capitalist states where individuals own properties and that it doesn't exist in socialist and communist countries where resources are vested mainly in the state. In my opinion, this assertion cannot be true because corruption finds its way into any human governance system, though in different forms. It exists in both market economies and nonmarket economies, and some governments, irrespective of the kind of economy, have been successful in curbing corruption. Thus, though the measures needed to bring corruption under control may differ depending on the type of governance systems a country chooses, the political will to curb corruption is critical in keeping corruption under control.

The severity of corruption is seen through the punishment meted out to corrupt officials in other countries. China, and North Korea impose death penalties if acts of corruption are established against a public official.[G1] The Philippines repealed the death penalty in 2006, but there are plans to have it restored. There are many other countries including[G1] Indonesia, Thailand, etc., that have capital punishment for acts of corruption, though it has been a while since corruption-related executions have been carried out. The severity of penalties for corrupt practices puts corruption among the category of serious crimes such as murder, drugs, human trafficking, among others. The severity of the punishment imposed by the state on corruption makes it quite clear that corruption is unhealthy for nation building and those involved in it are enemies of the country.

In a letter to an Anglican bishop, Mandell Creighton, in 1887, John Dalberg-Acton wrote that "power tends to corrupt, and absolute power corrupts absolutely. Great men are almost always bad men."[G2] There is no continent whose history has proven the words of John Acton to be accurate than the continent of Africa. Most often, African governments have been formed by genuine people with genuine intent to solve African problems. However, over a period, the power that was sought for the good of the people, corrupted officials of the government eventually erased most if not all the gains they made for the country. They remain in power, seek additional powers for themselves and remain in power even after they have lost touch of the issues facing the people they rule over, and send the country into deeper problems they came to correct. Cast your minds back to all the leaders of Africa. Pres. Robert Mugabe, for example, started very well but later plunged the economy he built into shambles.

Most African economies were doing well after independence, but corruption and mismanagement started to creep in, sending many countries on the path of destruction. The state no more became the concern of the politicians who administered the African economy. They became preoccupied with self rather the country. The success of individuals and families and friends was more important than the success of the state.

THE CASE OF DR CONGO

Though scholars have divergent views on corruption and its relationship with the economic development of a country, the case of the Democratic Republic of Congo (formerly called Zaire) provides a glimpse of the effects of corruption on a country's development. The Democratic Republic of Congo (DR Congo), which is Africa's third-largest country and arguably the wealthiest nation in African regarding minerals, forest resources, and fertile soils, is one of the countries that show how corruption can negatively affect a state. Zaire's former leader Mobutu Sese Seko is known to be one of the leaders who looted his country's coffers to become rich alongside Jean-Claude Duvalier of Haiti and Ferdinand Marcos of the Philippines.[G3] While countries like Singapore used about three decades to make a steady rise into middle and high-income countries, Mobutu used almost the same period to get rich cleverly. In criticizing corruption, Mobuto, on May 20, 1976, said, "If you want to steal, steal a little cleverly, in a nice way. Only if you steal so much as to become rich overnight, you will be caught."[G4] It is this attitude of the political leadership that sets the tone for corruption gain roots in the society. In the case of Zaire, the posture of Mobutu Sese Seko was undoubtedly going to take the eyes of the leadership of the country off the national agenda of building a strong nation to how they can benefit from the state. It was also going to inspire those who had avenues to steal resources of the country to seek clever ways of benefiting personally at the expense of the state. Eventually, the focus of the entire citizenry will center on how to make personal gains instead of how to solve the current and future problems of the state.

The distraction alone is a major danger to the economic development of the country aside diverting public funds away from where it is most needed into private accounts. The lesson of Mobutu, who saw his country, Zaire, sink further down into poverty while his assets increased in such a way that he made it into *Forbes* magazine's list of the world's dictators, should guide Africa's present and future leaders. Unfortunately, it appears Africa has not learned her lessons. Mobuto's example is becoming a good example for African leaders as they seem to show lackadaisical attitude when it comes to dealing with corruption. It is the kind of attitude that makes people who are in charge of resources engage in corrupt acts at the expense of those without the opportunity. It is this attitude that sends economies on a nosedive.

Nation building requires a sharp focus on policies and monitoring the effects of the policy on the citizenry. Focusing on any other issue including how to make money from the public coffers is a significant distraction in governance and can have dire consequences on the country. Mobutu's focus on stealing "a little cleverly" took his attention off his objective of seizing political power, and as a result, the nation's ports, roads, and the entire transportation network that needed expansion were left to fall into a state of disrepair. The healthcare system and public buildings were also severely affected. At the height of corruption, only the welfare of the individual is catered for, while the welfare of others becomes a burden, even the ones on whose shoulders you stand. In the end, Mobutu's and his family's well-being was the national priority. The country's infrastructure mattered less, and also, those who had the opportunity to steal cleverly laid hands on what they could get. Building a palace with a runway[G5] to serve Mobutu's family interest while the entire country's transportation system was in shambles is a misplaced priority fueled by unbridled corruption that set Zaire on the dark paths of economic decline.

Interestingly, from the time of Mobutu Sese Seko of Zaire, Africa, has seen the damaging effect of the corruption canker, and yet successive governments have not been able to put in place measures to curb it. There is no other reason for corruption to thrive especially in democratic states apart from the fact that governments and government officials benefit from the acts of corruption. Interestingly again, we have had opposition parties riding to power on the back of accusations of corruption of incumbent governments with strong words such as "zero tolerance for corruption" and other mantras. With such expressions, they gained the confidence of their citizens, who gave them the opportunity to rule and to curb the menace. Instead, they come into government for their officials to commit worse acts of corruption.

Ghana: Postindependent Corruption

I can only speculate why Nkrumah chose the path to socialism from hindsight information. The two most popular ideologies that ruled the world at the time Ghana had her independence were the capitalist ideology of the United States of America and the socialist ideology of Soviet Socialist Republic. The two nations were the world's superpowers of the time.

Ghana, upon becoming an independent country, was very rich in mineral resources. The country's sizeable illiterate population resulting from the skewed system of colonial education, which did not develop the capacity of the African, also posed a human resource challenge to the country's development. Additionally, as a new republic, there was the need to set up institutions and draft long lists of laws or regulations not only for the smooth administration of the country but also to control the resources of the new state. Given the above conditions, the choice of ideology or system of governance that was most likely to be successful in managing the competing individual interests to lay hold of the country's resources with the exit of the colonial government seemed to favor socialism.

The tendency for corruption to emerge and blossom in Ghana as a new republic was extremely high. I can only guess that Nkrumah foresaw that, unless the state becomes a purely nonmarket economy, where all the natural resources will be vested in and controlled by the state, the tendency for individuals to have interests in and seek to gain control over such abundant natural resources will be high. Thus, maybe to Nkrumah, it may be a good reason for the new country he and others have fought for to adopt a socialist form of government where the struggle to control the resources of the country will be minimized because individuals will not be allowed to own the country's resources. It was anticipated that if individuals' keen interests are not adequately controlled, people will want to find "ways and means," as it is called in Ghana, whether legal or illegal, to lay claim on the resources or the businesses around such resources of the state.

I further guess that Dr. Nkrumah's decision to take Ghana along the paths of socialism was a way to merge the state and the economy—to vest all resources in the state until the pillars of institutions that will provide the needed checks and balances can be erected to guard the behavior of the people in the new African country.

Controlling the economy was a way of keeping individuals from struggling for access to and possessing the nation's resources individually. Under his controlled government, only his appointees will be the ones with privileges and access to the nation's resources and the business around such resources and could, therefore, be corrupt. A further check on his appointees will be to set up "young pioneers" who will act as trained informants about his appointees, their families, and friends. Dr. Nkrumah tried to control the new economy by merging the legislative, executive, and judicial functions of the state, which had its downsides. Eventually, the excessive controls led to repression, which was the cause of his government's overthrow in the 1966 coup. I am tempted to believe that it was the attempt to curb corruption that got Dr. Nkrumah between a rock and a hard place, which led him to take decisions that came back to haunt him.

As I have indicated earlier, every system of human governance cannot be devoid of corruption. Kwame Nkrumah's government wasn't free of corruption either, but to take the country from a controlled environment to a more liberal economy was more likely to throw the economy into worse forms of corruption. Though Nkrumah may not have given his opponents a choice to make their contribution to the nation's government other than staging a coup d'état in 1966, the intervention of the military in the affairs of the state was a sure sign that Ghana has veered onto the undulating paths of corruption. And that was certainly going to slow down the nation's economic progress.

The sudden transition from a socialist state to a military regime made things worse, since governance and, by extension, the control of the nation's resources fell into the hands of individuals—soldiers. Imagine what will happen to individuals' uncontrolled access to the nation's precious natural resources. Military regimes set their own systems, and even if there were stronger institutions or systems to control corruption, they would have been immobilized because of the suspension of the constitution. The intervention of the military in the affairs of the state made corruption worse in Ghana. Even in properly executed transitions from nonmarket economy to market economy, some institutions, structures, and laws are set in place to check corruption, so a direct takeover from the military was undoubtedly going to create corruption problems. This is precisely what happened to Ghana.

The country returned to constitutional rule in 1969, but this time as a multiparty state with leanings to the capitalist ideology under Dr. Kofi Abrefa Busia. It must be situated in the context that the country had emerged from colonial rule in just a little over eight years ago. More so, Dr. Busia's predecessor and first president, Dr. Kwame Nkrumah, had turned to socialism, repression, and state-controlled economy where institutions and civil society including workers unions were suppressed. The military had also seized power and ruled with decrees for about three years.

With the above set of conditions, coupled with the fact that capitalism thrives on strong institutions to provide checks and balances, Busia's government was set for failure on the day it assumed office. This is because it was going to be very difficult for Dr. Busia to prepare and execute a proper transitional plan that will change the fundamentals of the economy from its colonial, socialist, military, and lack of institutions and civil society to support the new capitalist ideals that thrive on a different set of support systems. No wonder he admitted that the most challenging task facing his government at that time was how to eradicate bribery and corruption from the Ghanaian society. Sharing his frustrations about corruption, which was captured by the *Daily Graphic* on Monday, March 22, 1971, president Dr. Abrefa Busia said, "Bribery and corruption have eaten so deep into the very fabric of the society that when you put anybody in a position of trust, he or she uses that position to amass wealth."[G6] He further admitted that he would find it difficult to build a prosperous nation if the people continue to be selfish and dishonest.

Thus, without the underlying systems to control corruption and make it a disincentive for people to engage in it, Ghana's efforts at transferring her mineral wealth to her citizens were like building a superstructure without a foundation. It was just a matter of time; the building was inevitably going to come crashing down. Dr. Busia cannot be blamed because everyone who had the least opportunity to lay hands on any position or resource wanted it for himself, and the interest of the state was sacrificed for individual interests.

These individualistic paths on the part of Ghanaians as a result of the lack of proper mechanisms to deal with corruption exposed the steering wheel of governance to the military once again. On January 13, 1972, General I. K. Acheampong overthrew Dr. Busia's government, citing corruption as the primary motivation for him and his colleagues in the military to seize power. Ghana became trapped in a cycle of military and civilian government because, on the one hand, the military wasn't trained to govern, and they messed up the economy whenever they intervened to arrest a deteriorating condition. On the other hand, the civilian governments were also failing to achieve results because of lack of support systems to support constitutional rule. Within one decade, the periods between 1971 and 1981, Ghana had as many as six different civilian and military governments.

But it has been twenty-seven years since Ghana returned to constitutional rule in 1992. The country has set up all the institutions to fight corruption, and yet the issue of corruption remains endemic as Ghana is ranked eighty-one on the 2017 corruption perception index from transparency international. President Mills, when confronted with the evidence of an expose by an investigative journalist, Anas Aremeyaw Anas, went to the Tema Port and, after blasting the officials, gave indications that he was going to do more than to punish the perpetrators by instituting reforms. He never lived to achieve that dream as he died in office the following year. Almost every president of the Fourth Republic has been cited for one or two issues of corruption in his administration. The responses from the presidency have been the same—"Provide me with evidence of corruption, and I will deal with it." This is where the problem is because it is very difficult for ordinary citizens to prove acts of corruption. Individuals will only see signs or symptoms of corruption, but for acts of corruption to be established, it usually goes beyond what an ordinary person sees, hears, or does. Most often, strong institutions need to get involved for the truth to be established. The institutions in these cases have to be strong. Strong in word and letter. Strong enough to coerce every individual or organ of the state including the office of the president to provide specific information needed to establish an act of corruption.

Ghana has all the institutions it needs to fight corruption. Many other institutions were added in the Fourth Republic. The Commission on Human Rights and Administrative Justice (CHRAJ), the Economic and Organised Crime Office (EOCO), the Public Procurement Authority (PPA), and recently, the Office of Special Prosecutor are some of the institutions set up to deal with corruption. Whether these institutions are active in word and letter to successfully curb corrupt acts remains to be seen.

It can be observed that the youth of Africa are now beginning to get more and more involved in the politics of the continent. That is good news; Africa needs the youth in leadership positions because the future belongs to the youth, who are more likely live to see the future, but the bad news is that a more significant percentage of them are entering into politics with the wrong intentions. The intention is that it provides the short cut to personal wealth. These are not good signs in the continent's fight for economic emancipation and fight against corruption. It will only entrench corruption deeper into society. And in many countries with such rich, vast resources, it is just a matter of time that the fight for the control of current and emerging natural resources will break the nations through corruption. Rich Africa will be unable to transfer her riches to its inhabitants. In the end, the wealth of Africa will be transferred from Africa to a few Africans and many foreigners under the much-acclaimed democratic system of governance.

CORRUPTION AND ECONOMIC DEVELOPMENT

Corruption is a major hurdle that Africa has to clear on its path to prosperity and economic independence. Though others might have a different opinion and believe that Africa can continue to be corrupt and develop at the same time. There is no direct relationship between corruption and development or underdevelopment. Even in the academic circles, researchers differ on the issue of corruption. While some are of the view that it is a constraint on growth and development of a country, others are of the view that it "greases" the wheels of rigid laws and bureaucracy for better economic performance. Yet other researchers are of the view that corruption and development are not related in any way. It is, however, a fact that countries with lower corruption indices are developed economies that have higher per capita income, and the reverse is true.

The table below shows the corruption perception index and capital income of the top-ten and bottom-ten countries in the corruption index table.

COUNTRY	CPI RANK	CPI INDEX 2017	PER CAPITA INCOME
New Zealand	1	89	40917
Denmark	2	88	50541
Finland	3	85	45192
Norway	3	85	60978
Switzerland	3	85	65006
Singapore	6	84	93905
Sweden	6	84	50070
Canada	8	82	46378
Luxemburg	8	82	103662
Netherlands	8	82	52941

COUNTRY	CPI RANK	CPI INDEX 2017	PER CAPITA INCOME
Korea, North	171	17	1700
Equatorial Guinea	171	17	34865
Guinea Bissau	171	17	1700
Libya	171	17	9792
Sudan	175	16	4904
Yemen	175	16	2300
Afghanistan	177	15	1981
Syria	178	14	2900
South Sudan	179	12	1503
Somalia	180	9	N/A

SOURCE: *Transparency International*

Though one cannot easily project the issues of corruption as the main factor in Africa's underdevelopment because there are other influential factors, it cannot also be denied that corruption does not affect the development and progress of countries.

AFRICA'S CORRUPTION CHALLENGE

Everybody thinks he can't exist without taking some kind of favor or unapproved money to supplement his income. Judges take favors before pronouncing judgments, teachers take it and provide unearned scholarly marks, the civil servant and politicians take it before providing a government service. Whatever name we give to it, it cannot take away its legal name. The legal name is corruption, and it is illegal, and illegality can in no way become legal. The sad reality in our African society today is that we have come to a point where most people think that they can't survive in the days ahead if they don't get involved in the illegality. But the opposite is true. We can all survive if we all refuse to get involved.

The other side of the coin is that it is impossible for the one who doesn't engage in it to survive if others get involved and he chooses not to. So if one institution is corrupt and the other is not, it will affect the other negatively. The other being affected negatively also becomes corrupt and affects the other in such a way that the other also has to adapt to the corruption, creating the kind of corrupt society we have today.

What makes people put more attention on political corruption is that when those who lead get involved in corrupt acts, those who are not in privileged positions or don't have the opportunity to act corruptly begin to look for the least of opportunities to do so. In a matter of time, everyone, no matter which position he finds himself in, will also look for opportunities to act likewise, and eventually, corruption gets into every fiber of the society. To reverse the trend, there is a need to follow the same pattern. Leadership should refrain from it so that that they will have the moral right to punish those who get involved in it. Others will stop looking for opportunities to be corrupt because the leadership is firm and the consequences of being caught are too risky. Then eventually, nobody will want to engage in it.

The inability of many African governments to realize the objective of ending corruption largely stems from their failure to make efficient the systems to fight the problem. Making laws and setting up the institutions are not the panacea to the problem. Ghana, for example, has managed to pass and also propose many laws including the Public Procurement Act, which many civil society organizations were hopeful that it was going to curb the problem. To the disappointment of the hopefuls, that was not the case.

Sole sourcing procedures have been abused, and other manipulations to circumvent the laws of procurement still surface regularly. Obtaining government services such as passports is unduly delayed, while those who can part with something extra can quickly jump the bureaucratic hurdle. It is now the general perception that anything that delays at a processing office is because the officer working on it needs something. Thus, any government who will be able to reduce the timeline in the acquisition of public services will make a lot of progress in reducing corruption from the system.

CURBING CORRUPTION IN AFRICA

Research on corruption all over the world points out to the fact that corruption thrives when institutions are weak, and policies undermine competition and free trade. Failure to acknowledge the problem and the potentials of its existence is one of the primary reasons why the issue of corruption is endemic in Africa. As my old teacher used to say, understanding the problem is a major component in the process of solving the problem. It appears that Africa does not understand corruption, or if it does, it tries to deny its existence in the existing governing structures—I mean, the political circles, the traditional governance circles, religious governance circles, and the civil and public service governance structure.

I believe that the solution to curbing corruption in many African countries begins from the point of appreciation that issue of corruption exists and it has its deep roots within all the administrative and governance structures of our countries. It is only then that the entire country or society will begin to put in place systems that will help the country to confront the issue and deal with it. This reference point is the starting point of national development.

Once a country's leadership acknowledges the existence of the problem and understands the issue of corruption, it is only there and then that they get on the path toward exiting the dark hallways of corruption. It is only then that we can put in place laws, regulations, systems, attitudes, policies, and political will that will address the issues. Until then, the political rhetoric from our political leaders will mean nothing.

The attention and focus on political corruption are enormous. Focus on political corruption maybe another reason why Ghana as a country is unable to fight corruption because corruption is not confined to the political circles alone. I see political corruption as the branches or the leaves of the tree of corruption. If my assessment is true, then unless the stem, the taproot, and other roots that provide the nutrients to feed political corruption are dealt with, African countries are just scratching the face of the issue as we have been doing over six decades of political independence.

Corruption in Traditional Governance Systems

In chieftaincy or traditional leadership, chiefs do not sell lands, as lands are held in trust by the chiefs. From the days of old, when someone needed an area for building or other purposes, he had to inform the chief, who will give the person the land size required. The person had to provide a bottle of Schnapps (bottle of alcohol used for traditional prayers) as a symbol that he sought authority from the chief. No money was given to the chief except that the new trustee of the land (developer) pays for the land's related expenses such as survey, documentation, etc. Today, this is assumed to be the existing system, and it is understood that nobody pays for lands from chiefs, but it is far from the truth. Chiefs take huge sums of money from potential landowners, but these are not documented. The money paid is assumed to be symbols as the days of old, but huge sums of money exchange hands.

The challenge is that things change over time, and it is understandable that property developers cash in on the lands they use for the building, so there is nothing wrong with chiefs taking money from potential landowners. But why can't the chieftaincy system come out with proper ways of making sure the monies received by the chiefs are well receipted to aid in transparency and accountability? Receiving payments on the pretense that you have not taken any money is a form of corruption. The question that needs to be asked is, are those incomes taxed? Obviously no, because there are no indications that monies have been received, so how can the revenue authorities tax transactions that are assumed not to have taken place?

It is important to clarify the point that nobody is interested in taking away lands from anybody in whom that land has been vested. What needs to be done is that there should be transparency in every area of our governance system so that the institutions of state on which democracy thrives can provide the needed checks and balances.

Traditional leadership forms part of the governance system in Africa, and transparency in traditional leadership is a way of pushing political leadership from corruption. For the respect of the traditional authorities and the fear of incurring the wrath of the traditional authorities, many politicians have remained silent, but deep inside, they have concerns about the lack of transparency in traditional leadership. I remember a debate that erupted when a prominent chief in Ghana expressed concerns about corruption in the country. What got me more interested were the expressions that came from the people who were from different political divide. Majority of the people who interestingly were from different political divides were not happy about the lack of transparency with government funds that go to the chiefs in the form of royalties and private sources such as the sale of lands. Referring to the chiefs, one person said, "These people have no moral right to call politicians to order because they superintend over the worst forms of corruption."

As to what convinced him to make that comment aside from other comments that came up during the debate is unknown, but what can be inferred from what happened that day is, first, an influential voice that can help fight corruption is lost based on suspicions as there are subtle impressions that the traditional leadership is not transparent; and, second, the political leadership does not take the traditional leadership seriously when they voice out issues of corruption because they perceive the traditional leadership to be equal or worse when it comes to corruption.

It will, therefore, inure to the benefit of the entire Africans that the traditional leadership show leadership in fighting corruption by making sure that whatever seems to create doubts or transparency challenge among the people they rule over is made transparent and clear. In this way, they will be able to force the political leadership by their actions to follow suit because of the weight they carry in the African governance system.

Corruption in Public/Civil Service

Corruption finds itself also in the public and civil service in Africa. For example, in Ghana, through corruptions, nationals of other states can lay hands on the government of Ghana passports that are exclusively the rights of Ghanaian citizens. Most often, personnel serving in public offices see nothing wrong with taking some small money to provide documents to support other nationals' application for passports but failed to recognize that whatever is scarce is expensive, and the opposite is true. That is why you can travel to over two hundred countries and arrive at the airport of that country and apply for a visa on arrival because they know very well that once you hold a passport of that country, it is not likely to be a document obtained with fraud. The individual possessing it is considered a true son of America, and they know where to address any danger the individual might pose to the country. Can this be said of Ghana? No, I don't think so. That is why the number of countries that accepts Ghanaian passports without visa is decreasing each day. Because the only way other nations can protect their borders from criminals who have money to buy citizenships is to institute a visa, which they control its issuance, not on passport, which can easily get into the hands of those who have money.

A friend who used to work at the Ghana embassy in the Washington DC, USA, told me that once there was an issue of some Ghanaians who were arrested on minor offenses in the USA. The USA government, as part of the protocol, contacted the Ghana embassy to facilitate the deportation of the Ghanaians. The then ambassador, on probing further about the issue, realized that some of the people to be deported were not Ghanaians, and he suspected them to the Nigerians based on their accents. Recognizing the danger it might pose for the country if the people were not Ghanaians, he requested for a thorough investigation of their backgrounds, which dragged the deportation process over a long period to the displeasure of the American government. Eventually, some of the people admitted that they were indeed not Ghanaians.

The point I want to make is that corruption in the public service, this time not from the politicians, can get criminals of other countries' nationals into your country if they can get access to the country's security documents like a passport. Though the offenses they committed in the USA were mainly overstaying their visas, the situation would have been different if they were criminals. If they were criminals, it meant the country might have opened itself to importing criminals of other countries because corruption paved the way for other nationals to possess Ghanaian passports.

Interestingly, the process is ongoing. Ghanaians everywhere in the world is largely law-abiding. I have friends from other countries who got stuck to me when they realized I was a Ghanaian because of their previous experiences with other Ghanaians. These values are fast deteriorating because corruption has paved the way for other people who are not Ghanaians to be Ghanaians without going through the processes set out in our constitution. It is destroying the reputation of our country in the world, and it is getting worse with each passing day. I have traveled and seen other parts of the world since I was a child. But I have also seen how our reputation is going down the drain. Ghanaians used to have access to many countries of the world without visas if they were visiting for three months or less. Almost all these has been revoked, and a major contributing factor is that we have made our passports so cheap that friends of Ghana have to close their borders to us because they have to sieve out those who might be pretending to be Ghanaians. Just a few years ago, Ghanaians did not need a visa to go to the Bahamas, South Africa, Malaysia, etc. All these privileges have been revoked because of corruption.

Corruption in Religious Circles

It is hard to believe that religious organizations, who are thought of as holy and operate in strict adherence to moral and ethical principles that guard their beliefs, are now losing their credibility to perceptions of corruption. But it is understandable because John Acton's letter regarding corruption was directed to the Anglican leadership, reinforcing the point that corruption will creep into any human organization if conscious efforts are not made to contain it.

All across Africa, churches are springing up. Mosques and other beautiful religious buildings are also taking up spaces on the African landscape. The contrasting issue is that, while the springing up of religious organizations often leads to revivals and change of people's attitudes from bad to good, what is happening in Africa is opposite to the rivals experienced in other parts of the world. People are becoming more hateful and intolerant of each other, and evil and crime are rising with each passing day.

In South Africa, Nigeria, Ghana, Kenya, Zimbabwe, and many countries in Africa, many churches have been formed on the grounds of corruption. The differences that often result in the breakaway will be either the church administration is not being accountable to the people or a key figure in the church is not having his way regarding the finances of the church. But what is usually communicated to the general public is that God has called the pastor to form his church.

These answers usually end the conversation because if God is the referee, who can go to God and cross-check the facts?

The issue of pastors and religious leaders exploiting church members and diverting funds intended for religious activities are increasing with each passing day, and the news in newspapers across African are awash with such news. A major problem this is creating in Africa's fight against corruption is that another powerful voice that will speak to the issues is gradually being silenced because of such perceptions. If the people who can speak up are not speaking up or cannot command the needed recognition because their hands are equally dirty, then there is a great danger because the few will continue to make life bitter for the masses. The religious organizations, just like traditional authorities, have to come clean. They have to quickly find ways to dissipate the "corruption in religious circles" tag and stand up to play its role in corruption and nation building just as the early missionaries did when they came to Africa.

G1 Wikipedia (2019). Capital punishment by country. Accessed December 16, 2018 through https://en.wikipedia.org/wiki/Capital_punishment_by_country

G2 Acton, H. (1961). "Lord Acton". Chicago Review. 15 (1): 31–44. doi:10.2307/25293642.

G3 Denny, C. (2004). Suharto, Marcos and Mobutu head corruption table with $50bn scams. Accessed December 16, 2018 through https://www.theguardian.com/world/2004/mar/26/indonesia.philippines

G4 Richburg, K. B. (1991). Mobutu: A rich man in poor standing. Accessed December 16, 2018 through https://www.washingtonpost.com/archive/politics/1991/10/03/mobutu-a-rich-man-in-poor-standing/49e66628-3149-47b8-827f-159dff8ac1cd/?noredirect=on&utm_term=.c05200a56440

G5 Smith, S. (2015). President Mobutu's ruined jungle paradise, Gbadolite - in pictures. Accessed December 16, 2018 through https://www.theguardian.com/cities/gallery/2015/feb/10/inside-gbadolite-mobutu-ruined-jungle-city-in-pictures

G6 Daily Graphic, (1961). Bribery - Our Main Problem, says Busia. Monday, March 22, 1971

CHAPTER EIGHT

Technology And Data Management: The Necessary But Neglected Path

You know, I believe that technology is the great leveler. Technologies permits anybody to play. And in some ways, I think technology— it's not only a great tool for democratization, but it's a great tool for eliminating prejudice and advancing meritocracies.
—Carly Fiorina

For the past four decades, the focus of African countries has been to catch up with the world by closing the infrastructure gap. Many ports, schools, roads, railways, bridges, hospitals, and airports, among other physical infrastructure in Africa, are in deplorable conditions in the areas where they exist. Movements of people and goods are restricted by the poor infrastructure, and it comes as a higher cost to the people of Africa. The dire need of such basic infrastructure has forced many African countries to seek partnerships with institutions and countries that offer some hope in the midst of desperate situations. Many African countries are leveraging the natural resources to secure the much-needed funding to execute physical infrastructure projects. Some countries enter into good deals that bring some relief to citizens, while other funding deals raise concerns among concerned citizens in various countries.

It is essential to draw the attention of Africa and her leaders to the other side of the desperate attempts for physical infrastructural development. Yes, Africa needs money for its infrastructural development and needs to pursue that with all her energies. But what is the future telling Africa? What is going to be the future dynamics? The world is moving toward digital technology, and the earlier Africa gives equal attention to the development of her digital infrastructure, the better. Information or data is what is controlling the world now. Africa must not be found wanting in digital infrastructure in the next thirty to sixty years. The technological pursuits of the developed countries should not be neglected as Africa pursues its infrastructure development.

Maybe in the next few years, the roads Africa is sacrificing her resources to acquire may not be all that important. In a few months, the world is going to start using cars that fly. Is Africa developing the infrastructure or technologies that can support flying cars? Let Africa be watchful because by the time it finishes developing its physical infrastructure, there might be another wide infrastructure gap—the technological infrastructure gap. The roads might not be needed anymore, and Africa will be back to square one, where it has to sacrifice more resources to build something else. The way we perform daily tasks is changing at a faster rate, and maybe what will be needed will be digital networks that support the way to do business or even govern. Is Africa preparing for such a technological advancement?

Another question that is also worth asking is, is Africa developing her human resource in such a way that the human resource will have the needed skills to man the new technologies that are emerging? Or is Africa waiting for emerging technology to reach advanced stages before it starts training her people? Africa has had the experience of being late; that experience was a bitter one and needs not to be repeated. Africa has to get the future right. A famous Ghanaian adage says; "Ɛnyɛ atwee brɛ na yɛ yen kraman," which translates to "Dogs are reared long before the hunting season." Let Africa rear her dogs before the hunting season that is about to come. Africa's children need to be exposed to the technologies that are about to emerge. The only way to make that happen is to lay the foundations of digital infrastructure now. It is the sure way of giving Africa's children the opportunity to function in the global economy of the future effectively.

There is some infrastructure that Africa must have, but individual African countries might not have the financial or technical capability to build. In such situations, two or more countries must come together to make it a reality. This is where the African Union and other regional bodies should lead the effort. To achieve an essential feat in aviation technology, Britain, Germany, and France had to come together. The agreement that was signed in July 1967 by the three countries was "to take appropriate measures for the joint development and production of an Airbus."[H1] By 1970, the consortium Airbus company was born out of the necessity that Europe must play a leading role in the production of space, military, and commercial aircraft.

African countries have been making too much individual efforts in such a way that the power inherent in cooperation as evident by Airbus success is not being harnessed. When the Economic Community of West African States (ECOWAS) came together to create the West African gas pipeline, it helped Ghana in a way to solve her energy crises by providing relatively cheaper gas for the Takoradi Power Station at Aboadze.

The advancement or progress of any country is to the advantage of her continent and the entire world. Likewise, the advancement of every continent of the world is to the benefit of the world as a whole. Though others will share a different opinion, the fact remains—the world is one; a perfect world designed by the Creator for its various parts to work together but not against each other. Just as people at the time before David Ricardo developed his international trade theory of comparative advantage thought that by trading with each other, one country gains at the expense of the other, so it is in this in the world today. It is highly believed that the rise of one country or continent is a threat to the survival of the other.

That perception will change over time but maybe after the world has been wounded with the unnecessary competition. Though stationed in Toulouse, France, the Airbus company continues to bring immense benefits to the people of Europe and the entire world. The advancement of any country in Africa is not at the expense of or to the disadvantage of any other country in the world but rather to the advantage of other countries. Likewise, the progress of Africa will not be to the detriment of other continents. Neither is the advancement of America or Asia or Europe or Australia will be at the expense of any other continent.

If, for example, Cote D'Ivoire, the world's leading producer of cocoa, obtains a technology that helps it to produce more cocoa beans with lesser input, the benefit of that technology accrues to the country and the entire world. Because the world will have a lot of cocoa beans and the prices of cocoa beans will also come down, the world will enjoy tasty products made from cocoa at a lower price. Cocoa prices coming down will be seen in Cote D'Ivoire as a loss, but a second look at it portrays another picture. The technology reduces Cote D'Ivoire's cost of production, and this time, the country might not be getting more from higher prices but from higher quantities produced for the world.

Only greed, fear, and illusion of insecurity make people of one country think that the progress of another country is a threat to their well-being. The success of Japan was not to the disadvantage of the United States of America. The success of Singapore was not a threat to the success of Japan. The success of South Korea was not at the peril of Singapore. The success of China was not at the expense of South Korea. The success of the United Arab Emirates (UAE) was not a danger to China. And equally so, the success of emerging giants like India, Brazil, South Africa, and the rest will not pose a threat to any other country if the world will focus on the positivity of what these countries bring to the world.

Neglected Data

In the words of Peter Drucker, "What gets measured gets managed."[H2] Though several attributes cannot be measured in governance, the performance of a government is usually tied to statistics. For example, the rate of unemployment before and after a government comes to power, the crime rate, the inflation and interest rates, the literacy rate, the number of jobs created, and the GDP, among others. Governments' performances are heavily tied to statistics. However, African governments are unable to compile the statistics that help in making the day-to-day decisions. While often African countries track the population of their countries, other equally important statistics such as the percentage of the youthful population leaving their countries are not monitored for analysis and policy considerations. Africa has a vibrant youth. Over 60 percent of the 1.2 billion inhabitants is estimated to be below the age of twenty-five. The force behind this substantial human resource is gradually being dissipated through migration, but reliable data about it is neglected. Africa is losing this force to other continents including Europe, America, Australia, and, recently, Asia.

I did a rough calculation with my year groups at college and the university to get a sense of the gravity of the situation. With the help of a WhatsApp group created for our year group, I was able to track the whereabouts of all my colleagues. The college statistics indicated that 8 percent were dead, and 35 percent were still in the country pursuing the teaching profession or engaged in other sectors of the economy within the country. There were 57 percent who were out the country and spread across Australia, USA, Canada, and Europe.

The economics major class at the University of Ghana is worse as 68 percent were out of the country. I am tempted to attribute the high rate to the ease with which university students at that time had opportunities to travel on student programs like Au Pair, AISEC, and work-study programs. Many colleagues, in three years, had the opportunity to spend the long summer holidays in the UK and the USA on work-study visa or visitor's visas. All those who had that experience are no more in the country except for only one who recently returned to the country to set up consultancy, which he is currently combining with part-time lecturing.

A major concern that should be noted is that, while at the training college, all the students were receiving allowances as teacher trainees. Throughout the three years, the taxes of the people of Ghana were used to provide feeding, lodging, and stipend. If the statistics for my college year group is assumed and projected to be the national statistic, then 57 percent of that human resource trained with the people's taxes have now been moved out of the country because Ghana has not been able to provide them with what they need for a fulfilling life.

The Americans, through their Diversity Lottery Program, have taken a lot of Africa's skilled labor force. The Canadians, through their skilled immigration programs, have also absorbed a lot of Africa's skilled labor. Also, many Africans, through scholarships and further education, have become legal citizens in other countries contributing to the development of those countries, while the continent is in dire need of a skilled human resource. African governments hardly have any figures or numbers attached to the flight of her skilled workforce let alone drafted policies to bring them back or replace them. Should Africa continue to nurse talents with educational funding, which are mostly interest-laden borrowed funds from developed countries, to educate her citizens who will not grow to develop Africa but rather the countries where Africa took the loans?

I have met several of my colleague in the countries where they reside. Most of them have pursued further courses in their respective countries of residencies and are contributing immensely to the development of the economies where they reside. The interest to return to the country and help build Ghana with the additional knowledge they have acquired is high. What is left is the political and economic readiness of Ghana to receive their expertise. Some recounted that they had taken some initiatives to bring in new ideas to support the country's development, but they have been discouraged by the attitudes of the leadership in Ghana.

I recall a colleague telling me about how his hopes were dashed after traveling all the way from the United States of America to Ghana with the confidence that the self-financing technology that he believed was going to curb road accidents, car thefts, and other criminal activities was going to be well received by the Ghanaian authorities. To his surprise, the leadership he happened to meet was not interested, and he wondered what the interest of the government was.

He told me he needed a memorandum of understanding between the Ministry of Transport and the Ministry of Interior as the technology was to be jointly used by Driver and Vehicle Licensing Authority (DVLA) and the Motor Traffic and Transport Department (MTTD) of the Ghana Police Service. After several trips to Ghana, financed from his pocket, and all the bureaucracy involved in meeting with the ministers and officers involved, he never heard anything back from the ministries. He told me that one officer approached him at a later date and told him to offer "something" that will be given to officers who will work on his proposal, else the minister will not take him seriously. His response to that guy was "If the officers of the ministry will not take me seriously after all the pains I went through to customize the technology to suit the Ghanaian condition, then the ministry is not a serious ministry that has the development of the people at heart."

He further lamented that he knew funding would be a problem for the country, and that is why he developed the self-funding model for the Ghanaian situation, and if the country is not interested in a self-financing development project, then Ghana has a long way to go. According to him, the last time he followed up, he got to know that the minister was no more in office, and they couldn't even trace the whereabouts of his proposal. There was also no documentation that he has even been to the ministry or had submitted anything for consideration.

Ghana and many countries in Africa have not been able to obtain value from data properly, and that is becoming an obstacle to national development. Many African countries continue to spend huge amounts of money in obtaining different data at different times, while it could have been possible to derive accurate data from a single source with the use of appropriate data management technologies. Even among the same government, different government agencies spend time and money to acquire almost the same pieces of information for their operations, which at times leads to conflicting data and errors in government decision making.

To get into the specifics, let's take the example of Ghana. Few government agencies issues Ghana government identity cards. The passport is issued from the Ministry of Foreign Affairs, the Electoral Commission issues the voter identity card, the health insurance ID card by the National Health Insurance Authority, the Ghana national ID card by the National Identification Authority, and the driver's license from the Driver and Vehicle Licensing Authority (DVLA). All these agencies, at one time or the other, had to dispatch technical and human resources at huge costs across the various districts of the country to obtain information from citizens into a database for their operations. Some continue to keep lean staff in the districts or regions where they continue to collect additional data to support their services.

What happens is that, first, the money needed to obtain such biometric data at a point in time for the entire population is enormous. The National Identification Authority in 2018 estimated that it will require about $1.22 billion to compile and issue national identification cards.[H3] The amount is not only huge but also not available, so the government has to borrow from external sources with associated interest payments to compile and make use of the data. The Electoral Commission of Ghana also updates its data every two years to capture new voters and those who, for one reason or the other, were unable to get their names on the voter register. With effective data management, the country can get the job done with limited resources and can save the state a lot of money, which the country needs to attend to other equally critical developmental issues. It will be interesting if the country takes time to put together the cost of obtaining the data alone since Ghana's Fourth Republic in 1992 and compare that with the alternative of using better data management resources.

Second, aside from the money that goes to waste with the duplication of efforts by the various agencies, it makes it possible for people to obtain ID cards using different particulars. The danger the practice is exposing the country to is that people can easily change their identity by dropping his ID card with some set of details and going ahead to acquire a different genuine ID card with a different set of particulars. Except for the passport, which requires that the applicant provide a birth certificate as an attestation of citizenship and date of birth, applicants for other national ID cards have to give their date of birth voluntary. Voluntarily providing birth details and the opportunities presented at different times for citizens to get a form of identification make it easy for people to change whatever particulars they have as they move from one national identification authority to the other.

Even with birth certificates, the poor data management at the various hospitals, clinics, and birth centers creates the opportunity for people who want to abuse the system to do so. If a person reports a missing birth certificate and gets a police report to that effect, that person can go to the registrar of births to get a new birth certificate.

In that case, the person can change his date and place of birth by indicating that he was born at a hospital where he knows very well that the registrar of births cannot find any evidence to the contrary.

What these gaps in data management mean is that one can easily get a new identity with a different set of particulars and conceal whatever good or bad record that is associated with the old particulars. This environment is good breeding grounds for criminal activities. Though having even a perfect data management system does not guarantee that criminal activities will cease, the point is, if the status quo makes it very easy for people to conceal their identity after committing crimes, it will only motivate them to keep on committing more crimes. The case of driving in Ghana is discussed below to bring clarity to this point.

I am stressing the point because I know of a driver that has gotten into two serious accidents where, according to the police, his failure to adhere to simple road traffic procedures caused both accidents. Surprisingly, that guy did not go through any further training or receive any form of punishment, and yet he is still driving on the roads of Ghana. In each case, the guy was driving again in less one month of causing the accident. How many more people should he kill before the country stops him? If it were to be in any other country outside Africa, he would have lost his freedom to drive for many years even after the first accident. He would have undertaken some extra driving courses before getting his license back. In that case, the second accident wouldn't have happened, and somebody whom he has sent to an early grave because of his negligence may still be alive and contributing to the development of the country.

The statistics regarding accidents in Ghana is alarming. According to the information released by the Motor Traffic and Transport Department (MTTD) and the National Road Safety Commission (NRSC) of Ghana, there was 1,212 death recorded in the first six months of the year 2018 through road accidents, while 6,698 persons were injured in the accidents.[H4] The figures round up to seven people die each day through road accidents. Meanwhile, most of these accidents could be avoided if the drivers were to adhere to road safety instructions and obey simple traffic rules such as keeping within the specified speed limit and overtaking at areas marked as such. I don't know what disease or illness can take such number of lives daily, but I don't think Ebola at its peak was claiming lives as much as accidents do in Ghana. More importantly, few of these drivers die in the fatalities. They recover from their minor injuries and are ready to hit the road again to do what they do best, bringing carnage on the roads and sending people to their early graves.

With better data management systems, Africa can sift the good drivers from the bad one. It is possible to have the licenses of all drivers captured in an interactive database, which, with the help of the appropriate device made available to the motor traffic officers, they can record on the spot the traffic offenses of drivers. A driver caught will have the option to challenge the officer's accusation in court when he thinks otherwise. With such a system in place, no matter where a driver's name or license number is inserted in the device, all the driver's previous traffic offenses will pop up. Transport companies and companies that use a lot of vehicles for their operations can benefit immensely because they can easily contact the police or driver licensing authorities for a background check on a potential driver's driving behavior. The companies will be saved from handing over their expensive equipment to careless drivers who drink, drive, and, above all, put their lives and the lives of others in danger on the roads of Africa.

The countries in Africa continue to record high levels of avoidable accidents on their roads. Whenever such accidents happen, especially the fatal ones, people lament and ask questions, but the answers are right before our eyes. People are driving anyhow because they know that no matter what offense they commit, data management systems will not be able to trace their previous behaviors, and their driving permit cannot be taken away from them because the current system does not have that option. Until African countries begin to use technologies and better data management solutions in controlling the behaviors of their citizens, the actions of a few non-law-abiding citizens will continue to make life unbearable for the many law-abiding citizens. People will continue to drive dangerously because we allow them to, and with such behaviors, there is no way Africa can put an end to the carnage on the roads.

Let me share an experience in Japan that has been on my mind for a long time. I friend who emigrated from the Middle East to Japan some years ago shared with me his driving experience in Japan. Once he was asked by a friend to bring his truck from Tokyo since he was going to Tokyo for a meeting the following day. He couldn't say no to the friend who had been helpful to him in many ways some time past, so he answered in the affirmative, though his license did not permit him to drive bigger trucks. While on his way back with the truck, he was stopped by the police on a random check. They tested him for alcohol, and it was negative, and everything about the vehicle was fine. Finally, they asked him for his license, which he reluctantly provided. They realized he had permission to drive smaller cars, but he wasn't allowed to drive trucks.

Meanwhile, in his car business yard, everybody knew he has been using trucks to pull vehicles and taking them from one place to the other for many years. The police impounded the vehicle, and for the offense of driving a vehicle he was not permitted to, he was banned from driving all forms of vehicles for five years in Japan. After completing his five-year ban, he had to take a course in driving at his own expense and pass the course before he got his driver's license back.

In my opinion, the Japanese are people who know the value of human life and who will leave nothing to chance when it comes to protecting citizens. My friend was not involved in an accident; neither was he caught committing any traffic offense. He was stopped at a random police check, and he faced the law. To the Japanese, if you can drive, it doesn't mean you can drive any vehicle; but to the African, even those without license can drive. If you are not qualified to drive a particular vehicle, you must not do so because you might not know what to do in times of an emergency, and someone might get maimed or killed for your lack of skill.

Africa's inability to enforce traffic regulations may be the reason why, according to *Wikipedia*, the highest death tolls tend to be in African countries. What Africa is failing to grasp is that in the minds of people are noble ideas that the world is yet to experience. As Martin Luther King Jr. puts it, "The lightning makes no sound until it strikes."[H5] Thoughts in the minds of people are not seen until they begin to put them into actions at the appropriate time. The noble ideas of many people African are sent to the grave because a deadly weapon (car) was left in the hands of a careless individual who took a heavy dose of alcohol and decided to drive while he is not permitted to do so.

Back to my friend's story, I asked him if he drove during his five-year ban. He said no, and when I asked why, he replied, "I will go to jail if they catch me driving even if I don't get in an accident or commit any traffic offense." Our conversation about road traffic issues was quite lengthy because we were driving in his car from Nagoya to Osaka. He recounted the business he missed because of the ban, not forgetting the inconvenience he had to go through as he could not go anywhere without looking for somebody to drive him.

I narrated the different situation we had in Ghana and, by extension, in many other African countries. Innocent people are slaughtered by people who drive without a driving permit. There are those who operate buses and eighteen-wheelers on our roads with a license for driving smaller cars. African has the laws to keep the people safe, but the laws are just in the books. So far as Africa makes it easy for people who don't want to obey the rules to operate, the more those who live by the laws in the society suffer.

THE WAY FORWARD

In an attempt to provide services to the people in a country, a government has to deal with a lot of data. There are so many data about citizens, corporate citizens, governmental organizations, civil society organizations, institutions, etc., that the government needs to gather and process for the good of her citizens. Advancement in knowledge and information technology has made this task very easy. Unfortunately, not all countries have taken advantage of the progress in information technology and data management to harness the power of effective data management to serve her citizens. According to NGDATA, the term "data management" is an "administrative process that includes acquiring, validating, storing, protecting, and processing required data to ensure the accessibility, reliability, and timeliness of the data for its users."[H6]

The way forward for Africa is to adopt the appropriate technologies in public administration to improve efficiency and effectiveness. Productivity in African countries will increase when there is efficiency in the provision of public services. Almost all enterprises will have to deal with the government at one time or the other in their operations. Common complaints from the business community are the bureaucracy in dealing with the government. Efficiency in the provision of government services is therefore critical to driving the efficiencies of the whole economies up. Technology and data management hold the key in such areas.

To address the issue of my friend whose proposal to the ministry got missing, I am sure that in as much as departments continue to log letters manually in foolscap notebooks, such errors will keep occurring. However, I am sure a simple software that records letters, proposal, etc., as they are received can easily solve that problem. For example, the secretary gets a letter and date its. The receipt of the letter is logged into the computer, which automatically prompts the next officer who will work on the document. The next officers will also log in the time they picked the material from the secretary and the action taken on it. The following officers will make similar inputs in the software until all activities on the document are complete. With such arrangements in place, people who are paid with the citizens' taxes will be productive because it will expose the officer who dropped the ball. For instance, if the secretary forwarded a document to the chief director, and the chief director sent it to the technical team, who acknowledged receipt but failed to act on it, it will identify the person who is letting that ministry down with nonperformance. In that way too, citizens accessing public services will be able to know the status of their application with ease.

In the area of security, Kenya has shown the way. To curb crime, Kenyan Police Service has established the traffic command and control center in Nairobi.[H7] It is an advanced information technology-driven system to check criminal activities in the nation's capital. The system has installed over 1,800 CCTV cameras at strategic places to monitor the activities of criminals. It includes a call dispatch center that handles over forty thousand calls in a day.

Africa also needs to empower her research institutions. Africa has neglected its research institutions and yet wants to achieve economic development. Countries in different parts of the world achieved economic growth using their peculiar development models. Those models, if not developed by the countries themselves, were scrutinized for its adaptability in the context of the country's specific situations before being adopted. Africa is fascinated by the Chinese economic development model. The continent is also intrigued by the Singapore example, the Japanese case, the Korean example, the UAE example, etc. Presidents of African countries on their visits to these countries are eager to adopt the models of the economic success of other countries. There is nothing wrong with learning from others since it is a waste of effort to reinvent the wheel. However, Africa has to realize that every model has different variables. The behavior of the individual variables in the equation is what determines the outcome of the model. Though there may be similar variables, there will always be at least a single variable that needs to be identified for a particular country.

The research institutions of Africa, from archeology to zoology, needs to be resourced to participate in the effort to come out with ideas for achieving the economic development of the countries. Concerted national efforts should be made to coordinate the tiny pieces of knowledge emerging from research institutions. These pieces of information need to be joined together by professionals for policy formulation implementation and devoid of political differences.

Moreover, as indicated earlier, Africa has to build its digital infrastructure. The Internet connectivity in many African countries is limited. Currently, Africa accounts for 21.8 percent of global Internet users—the lowest in all the regions of the world. The statistics need to be changed, and Africa must have the appropriate technology to support her development at all costs.

From General Packet Radio Service (GPRS), mobile Internet has moved to the second generation (2G) and now to the fourth generation (4G), which is ten times faster than the 3G. Now, 5G, which is expected to introduce the "Internet of things," is almost here, and who knows what will come next? It is anticipated that in 5G, not only will our phones and tablets be connected to the Internet but also fridges, houses, cars, etc., will be hooked on the Internet. Phone manufacturers including Samsung, OnePlus, and Huawei will, from anytime this year, make available 5G phones. Many cities in advanced countries are bracing up for the Internet revolution. Companies manufacturing fridges, washing machines, etc., are also preparing for the rollout of 5G.

High-tech devices are on the increase. Companies manufacturing these high-tech devices are finding ways to connect them to the Internet to allow its control from the handheld devices. The devices will surely require extensive bandwidths. The important thing to observe here is that the advanced countries are forging to make mobile Internet connectivity as fast as possible and making it more able to transmit more data within the shortest possible time. What can be done with these advancements is unimaginable for now, but what is imaginable is an Africa that is ready for the unthinkable. When 5G becomes fully functional, it will revolutionize everything. There is no way education, health care, transportation, and the method of doing business will ever be the same. Policy in Africa should make Africa's human resource and infrastructure ready for the anticipated changes before it is too late for the continent.

This is the era of drones. Some companies like Inkonova in Sweden are now building drones that can create and scan 3D maps of previously inaccessible areas. Such advanced aerial and underground technology has many implications for resource-rich Africa. It is possible for other companies or countries to estimate the true value of a mineral deposit by drones and negotiate a lesser price for the concession that African government may accept because they are not privy to the information they have about the resource.

Every area of man is being affected by the advancement in technology—agriculture, medicine, entertainment, transportation—in such a way that it is anticipated that there will soon be a four-day working week. If Africa's research institutions do not have the comparative advantage in coming out with new technologies because they are underfunded, they should be able to monitor advancement in technology in different sectors and help shape policy in a direction where Africans will be in the position to embrace new technologies.

Africa should also adopt technology and data management and avoid heavily relying on cash for economic transactions. African countries spend huge amounts of money on printing new currencies, while the point-of-sale terminals can be used to reduce such costs and save people from armed robbery attacks. In Ghana, for example, several people have become victims of armed robbery attacks because they carried lots of cash with them. An introduction of an effective electronic system should reduce incidents of robbery or confine such attacks to financial institutions that have security systems to foil robbery attacks. The challenge might be the levels of illiteracy, but the success of mobile money in Ghana, Nigeria, Kenya, Tanzania, and a host of other African countries shows that is possible to roll out an efficient electronic cash system in African countries. Ghana initiated the "e-switch" some years ago. Though patronage is still relatively low, it is a step in the right direction, and lessons can be gleaned from it to perfect it.

H1 Simons, G. (2014). The Airbus A380: A History. Pen & Sword Books

H2 Drucker, P. (2012). The practice of management. Routledge.

H3 Ghanaweb (2018) Ghana to spend 1.22 billion dollars on national ID cards – NIA reveals. Accessed December 16, 2018 through https://www.ghanaweb.com/GhanaHomePage/NewsArchive/Ghana-to-spend-1-22-billion-dollars-on-national-ID-cards-NIA-reveals-654375

H4 The Finder (2018). Rising recklessness on roads 1,212 Killed in 6 months. Accessed December 18, 2018 through https://www.thefinderonline.com/news/item/13549-rising-recklessness-on-roads-1-212-killed-in-6-months

H5 Quoteswise (2018). 411 Martin Luther King Jr Quotes. Accessed December 18, 2018 through http://www.quoteswise.com/martin-luther-king-jr-quotes-8.html

H6 Galetto, M. (2016). What is Data Management? Accessed December 18, 2018 through https://www.ngdata.com/what-is-data-management/

H7 Ombati, C. (2013). Uhuru unveils new traffic command centre. Accessed December 20, 2018 through https://www.standardmedia.co.ke/article/2000099808/uhuru-unveils-new-traffic-command-centre

CHAPTER NINE

Donor Partners: The Borrowed Path

Rarely do we find men who willingly engage in hard, solid thinking. There is an almost universal quest for easy answers and half-baked solutions. Nothing pains some people more than having to think.
—Martin Luther King Jr.

African countries have centered their economic development on countries and institutions they refer to as donor partners or development partners. Many will doubt this assertion, but as the saying goes, actions speak louder than words. A country with an appetite for economic development cannot leave half of her budgetary commitments to chance (aid) and expect to achieve sustained growth, but that is what has been happening year after year in most African countries for decades. According to the Institute for Security Studies (ISS), "Without aid, government revenues in Africa's 27 low-income countries would be 10% of GDP instead of 21%."[J1]

While many African countries depend on the assistance of donor partners to provide basic health services, scholars have divergent views on how aid contributes to the socioeconomic development of third world countries. There are arguments for and arguments against using development assistance to pursue economic development objectives in developing countries. Karikari, for example, believes that dependence on foreign aid "induces a lazy, slavish, dependent mentality and culture across society—from governments to villagers." But Egbert de Vries thinks that a certain amount of aid can achieve "a modest increase in the standard of living of the poorer."[J2]

Despite the unconfirmed correlation between aid and development, Africa's donor partners, maybe with the best of intentions, have been giving development assistance to Africans for decades. African countries and friends of Africa have called for more funding up to 0.7 of gross national income (GNI) of the donor partners and a better approach to make aid more effective. Norway, UK, and three other countries have exceeded the 0.7 percent aid target, and year after year, conferences on aid effectiveness have been held between African countries and their donor partners, but there seems to be little or no change in Africa.

Something should strike a chord in Africa. Are the terms "development partners" and "donor partners" appropriate? At times, people carve out socially accepted names or terms to avoid embarrassment and shame. For instance, instead of calling countries "undeveloped countries," the term "developing countries" is carved to indicate that those countries are in transition from undeveloped to developed status, but the process – by the current trend - may take forever for many countries. Maybe if the people of Africa start calling their relationships with other countries the way it should be called, it will arouse a wave of "positive" anger within Africans to make efforts to address the imbalance.

If Africa refers to foreign donors as donor partners because we are referring to them based on their relationship to us as donors, then how should the people in developed countries call us? Apparently, "receiver partners." But that is not how they call us. They call us "receiver countries" because there cannot be any meaningful partnership when the role of one partner is always to take something away from the partner and the other partner's position is to give all the time. That partnership, if it exist, will be a weak partnership. In science, a tree that feeds on the other to survive is called a parasite. The fact remains that those benevolent countries giving to Africa on humanitarian or whatever grounds are "donor countries" and cannot be Africa's "donor partners."

Others may disagree by raising issues about donor countries exploiting Africa in the past to justify that donors benefited at one time and so it is Africa's turn to benefit from the partnership. I find this ridiculous for two main reasons: First, if donor funding is to compensate for the exploitation of Africa, then donor countries like Sweden, Norway, and Japan who were not involved in colonialism and the scramble for Africa should have no business giving to African countries. Second, the Jews in Israel have gone through several ordeals and have arguably suffered more than what Africans have experienced. However, they did not focus on getting help from those who hurt them. They started doing something to change the situation.

THE IMF AND THE WORLD BANK

The IMF and the World Bank are institutions that are familiar to many African countries. In the 1980s and 1990s, the institutions, in response to Africa's rising debts, offered loans and economic and political reforms to African countries. The IMF expected African countries to graduate from their programs; instead, after over four decades of intervention, African countries have been hooked to it as more and more African countries continue to seek emergency loans from the international institution.

Latest IMF programs are no more giving too many loans; it is aimed at helping Africans develop strong economic foundations. IMF monitors the policies of African countries and approve or disapprove local policies. If the plans are approved by the IMF, then development banks and other donor countries come in to give more loans; but if they fail to get the IMF seal, then some possible funders or donors hold their funds.

However, the policies of the World Bank and the IMF have always been the same—generate enough local revenue, cut spending, privatize state corporations, and liberalize trade, among others. In essence, IMF/World Bank programs are just a micromanagement of economies of countries that are in distress. Interestingly, they appear stricter for developing economies like African countries than for the more wealthy states. The simple point I want to raise is that IMF comes in when countries are in distress or have some financial difficulty. I have yet to see a country who is doing well economically seek help from the IMF. If IMF/World Bank programs are for countries in distress, it will be a disaster for Africa to continually depend on the IMF to micromanage their economies toward greater prosperity. Ideally, once a country emerges from a distress situation, that country should pursue a more ambitious program aimed at securing economic prosperity for her citizenry. This can only happen when African governments and her people are ready to take the bull by the horns and do something extraordinary than the dependence on donor aid from other countries and the grants and loans from the IMF/World Bank. African countries have walked on these paths before. They have a long history with the World Bank and the IMF. To be fair to the institution, some of the loans and aid programs yielded some gains. Ghana's economy was saved from total collapse with the implementation of the structural adjustment program in the '80s, which turned the negative growth rates to a positive one.

The Ghana experience is good, but this is where Africa is failing. Again, it comes back to purpose. So the question that African countries need to ask is, what is the goal of IMF/World Bank programs? As mentioned earlier, they are programs that save economies from total collapse. The programs are not prosperity programs, and countries seeking growth and prosperity programs should look for where they will get them. Anyone who will go to the hair salon for a health check is instead asking for some more severe health problems. Countries that have achieved prosperity chartered their paths and walked on them. They did not do so by turning to the IMF/World Bank. They designed their growth and prosperity programs and made sure that those policies took them to where they wanted to be. Africa needs to pursue programs that are intended and designed for economic prosperity.

Africa needs to wean itself from the Bretton Woods Institutions. It will mean that African countries should have a long-term plan to graduate from the "IMF/World Bank austerity university programs." Not every African country can graduate at the same time, but others who have achieved some level of stability should embark on ambitious prosperity programs that will create a solid foundation for the economies before unexpected economic shocks force them to reenroll on the programs of the Bretton Woods Institutions.

DONOR OR IMF/WORLD BANK DEPENDENCY

What Africa has failed to learn from her dealings with the World Bank and IMF is that the institutions do not have all the answers and their well-intentioned policies can fail. Like all policies, the omission or addition of a factor in the fundamental model can lead to different results. Commenting on IMF's program-backed bailout to Greece, which injected a lot of funds into the economy without the expected turnaround of the economy, Olivier Blanchard pointed out in 2013 that the IMF-approved policy of cutting government expenditure did more harm than envisaged.[J3]

Africa has to learn that after four decades of IMF/World bank policies without the needed economic independence, it must try other methods. At least Africa can learn the lesson of Argentina. Argentina rejected the IMF and pursued its programs for over a decade. Pres. Néstor Kirchner of Argentina, in severing ties with the IMF in 2005, said the move was to give Argentina more autonomy to pursue its economic policy.[J4] Though the Argentine financial crises of 2018 forced the country to open talks with the institution, this time, the IMF has pledged to provide Argentina with US$57.1 billion, the largest funding that the IMF has offered to any country.[J4] Though Argentina is back to the IMF, their commitment to pursue their home-grown programs in 2005 was worth it because it took them to a point where if they will need the IMF's support, it will be a bigger bailout fund, unlike the peanuts handed to Ghana in 2016, which is less than $1 billion.[J5]

Even when African countries economies are hit by crises like Argentina in their attempts at becoming prosperous, it will be worth giving it a try because there will always be options. Africa can still learn from the countries that have gone ahead of us. Every economy will face some form of financial crises at one point in time, but there are many ways to get an economy in distress back on track. During the Asian crises, Malaysia, Thailand, and Indonesia were some of the Asian economies that were hard hit.

Thailand and Indonesia chose the path of the IMF, while Malaysia refused and stayed on its path toward recovery. In the end, all the countries recovered from the crises. But the statistics are fascinating to observe. In 1998, Thailand's economy, which chose the IMF/World Bank path, shrunk by 10.5 percent and recovered to 4.4 percent in 1999. Indonesia, who also chose the way of the IMF, shrunk by 13.1 percent in 1998 and rebounded to 0.8 percent the following year. Malaysia, who rejected the IMF/World Bank program and chose her own economic path (capital controls), had her economy shrinking by 7.4 percent in 1998 and recovered to 6.1 percent in 1999.[16]

Standing alone or blazing a trail is not easy, and it is not meant for the faint-hearted. It has its fears and uncertainties. Lesser known paths are precarious and can have a devastating effect on those that walk on them. But it can also be rewarding. African countries need to "take the first step in faith. You don't have to see the whole staircase, just take the first step," Martin Luther King Jr. suggests. Africa has had the interventions of the IMF and World Bank for a long time. Countries enroll in one program, get out of it, and, within a few years, are back on another kind of program that is aimed at addressing the same fundamental issues. The economies of African countries are still struggling economically. Africa has a choice—to continue to be in its present course where its economies will eventually go on its knees for IMF and World Bank to resuscitate and then pursue similar policies for the country to get back on its knees again or to take the bold step and try something new like the Argentines did.

Maybe the ambitious programs might not end well, but who knows, it might end well just as it happened to Malaysia in 1999. The choice remains with African leaders, Africa's civil society, and the African people. Africa chose its path when it became independent. It will be hard to believe that the founders of independent African states in the likes of Kwame Nkrumah of Ghana, Patrice Lumumba of the Congo, Gamal Abdel Nasser Hussein of Egypt, Julius Kambarage Nyerere of Tanzania, Ahmed Sékou Touré of Guinea, Modibo Keita of Mali, Félix Houphouet-Boigny of Cote d'Ivoire, Nnamdi Azikiwe of Nigeria, Jomo Kenyatta of Kenya, Kenneth Kaunda of Zambia, Samora Machel of Mozambique, and Nelson Mandela of South Africa (just to name a few), at the time of the independence struggle, had in mind that it was going to the IMF/World Bank or other countries that were going to manage their economies with unknown economic programs.

The path of the IMF/World Bank is not Africa's path. It is an unknown path. The way of the IMF, the World Bank, or any other foreign country should be seen as a detour that only becomes necessary because of some roadblocks that we or others placed on the paths to economic independence. Africa cannot continue on the bypass because that road leads to nowhere. Africa needs to get back on its path toward prosperity. Africa's path to economic success seems dangerous and threatening. Indeed, it may be bumpy and discouraging to tread on, but using the seemingly smoother roads of others to get to your destination is a lazy way out, if even that is possible. The end of Africa's path to economic independence is in sight, but only the courageous nations will get there because it requires greater sacrifices than the sacrifices that gave the continent its political autonomy.

Take the case of Mozambique, in a 2007 report by Action Aid International titled "The IMF's Policy Support Instrument: Expanded Fiscal Space or Continued Belt-Tightening?" the government of Mozambique was pursuing an educational program that sought to reduce her teacher-pupil ratio from 1:72 to 1:40. The IMF program superimposed an inflation reduction targeted program that called for the country to reduce government spending to achieve the inflation target.[17] It meant that the Mozambique government would not be able to hire more teachers to curb overcrowding and improve the quality of education in the schools because of the strict policy. The country had to follow a plan that was apparently in conflict with the present and future aspirations of the country. Can such counterproductive policies help the countries of Africa to achieve their prosperity dreams if the future quality human resource, which it desperately needs, is sacrificed for a short-term inflationary target that can be achieved through other policies? With such a policy that undermines the future of the country, Mozambique can be sure that the next generation of Mozambiquans will be less trained than her colleagues in other countries. How can the next generation negotiate better international deals for their country when her human resource capacity was weakened as a result of following counterproductive policies that were forced down on her throat?

AFRICA'S NEW DEVELOPMENT PARTNERS

Though many African countries are still on IMF/World Bank programs, many African countries are looking elsewhere for the grants and loans they used to secure from the Bretton Woods Institutions. Africa now has new emerging partners. Key among them are the seven countries of Brazil, China, India, Korea, Malaysia, Russia, and Turkey, which are referred to as New Emerging Partners (NEP7). The NEP7 states have increased aid to African countries. They are helping to build Africa's infrastructure, though most times, it is an indirect way of providing jobs for companies in their respective countries with the infrastructure loans they extend to Africa. What is getting Africa hooked to their new development partners is that they operate in a different framework. Whether the framework is in the interest of the citizens of Africa remains to be seen. What is certain is that it is more convenient and appealing to the leadership of Africa. Unlike the strict labor, environmental, and transparency rules that govern the operations of Development Assistance Committee (DAC) economies and firms' operations in Africa, NEP7 companies usually have no restrictions.

What Africa has to understand is the fundamental truth that the primary focus these countries is not to help Africa to develop, though they often champion projects that are of strategic importance to the economies of African countries. Their primary objective remains to serve the people of their countries. In this era of nationalism, where the United States president can boldly call himself a nationalist, Africa cannot afford to entrust the hopes and aspirations of her citizenry in the hands of other countries and institutions.

Countries are becoming more inward looking. Policies that are favorable to the citizens of their countries are given priority, damning the consequences of such policies on other countries. Countries apply tariffs to goods that they perceive are helping build other economies at the expense of theirs. The year 2018 saw the imposition of tariffs on some Chinese products by the United States government with China reciprocating the "kind gesture." The 2018 U.S.-China trade war and the renegotiation of the NAFTA deal are indications of how introspective countries have become. What was at stake was the well-being of the nationals of these countries. The spiral effects of their actions on the global economy mean less than the welfare of their citizens. Only African countries and the leadership still believe that other countries will take a decision or put in place policies that will be in Africa's interest.

In the midst of the U.S.-China trade war, the words of the Chinese leader continue to impress me. As reported by the CNBC, the Chinese leader Xi Jinping called on the good people of China to "stay on course" and that China has the "right to pursue its path going forward."[38] Not all leaders lead; some leaders lead, while others follow. Such words are some of the distinguishing features between leaders who lead and leaders who follow. China has followed its path; it has made serious sacrifices and attained economic success to the surprise of many. It has other paths to follow—the path of playing a dominant role in global affairs. And so long as they stay on their path, the end is in sight. China is not preventing any other nation from having ambitious programs. Others can choose to rest, others can abandon their journeys, but for the people of China, they are following their dreams. The world can say whatever they like about their country, about their leaders, about their people, about their chosen paths, about their economy, about their governance systems, about their governing party, about their human rights records, about their laws, about their currency, about their policies toward Africa, even about whom they choose as friends. That will not amount to anything provided the people of China know where they are going; what matters most is their ability to "stay on course."

AFRICA'S RELIANCE ON DEVELOPMENT PARTNERS

Many people in Africa and, sadly, many policymakers and ministers are of the view that governments, agencies, and institutions of other countries have an obligation toward us and that they exist to protect the interest of Africa. I once heard a minister say that "as soon as we discussed with them, they saw that it was good, so they were ready to fund it." When I heard this coming from a minister of state, my heart ached because I knew that the future of Africa was gone because there are no long-term plans to liberate Africa from dependency. The fact that a policy receives funding does not mean it is a good policy.

African governments should not be in a hurry to institute programs with no sustainable funding just because they are excellent ideas. Political campaign messages should not be based on anticipated support from donors. Ghana wanted to provide a meal for every child in primary school popularly known as the School Feeding Program. The development partners pledged their support for the program. Somewhere down the line, they withdrew their support, which led to months of arrears owed by the government to the caterers who prefinanced the feeding of the schoolchildren. Eventually, it became a government of Ghana–funded program, which had to be exclusively funded from government's resources.

Africans should also stop demanding "free" things that push their leaders to make "imaginary" promises. It is important for Africans to note that, aside from the gifts of nature, the prices placed on things are almost equal to the value of the efforts put into producing and making it available to the consumer. Whatever the government will give freely to her citizens has to be paid by the government. The more freebies Africans request from governments, the more the government has to raise revenues, and in most African countries, increasing revenue is synonymous with raising taxes.

At times, it is confusing to see governments of African countries who are unable to function if they don't receive support from donors promising to deliver additional capital-intensive government services for free. Interestingly too, the citizens become excited with such juicy promises and vote them in only to be hit with the reality that the government has to "squeeze" them further in taxes to fund that "promised" project. It is time for the leadership of Africa to be truthful to the citizens. It is also about time that the citizens appreciate and accept the reality that governments face a lot of challenges in revenue receipts, revenue leakages, and adequate funding to cater for the citizenry. It is only when the two parties come to this agreement can the leadership build the needed synergies to get things done for the people.

As it stands now, many African governments are afraid to give the actual state of the economy to their citizens. Statistics and analysis that will paint the actual state of affairs are at times massaged before being communicated to the general public. The populace continues to have false hopes with the anticipation that very soon, things are going to change for the better, until they hear the government that they know is doing well being rushed to the intensive care unit of the World Bank or the IMF.

The civil society, the media, and the entire African populace should begin to analyze the policies of politicians. They should not just be interested in the sweet talks, but instead they should be interested in how those policies and programs are going to be funded and implemented. More importantly, what sacrifices need to be made to achieve that objective. The freebies will bring undue pressures on the economy and keep the leaders to repeatedly run to those whose interest might not be in the interest of Africa, who might want to exploit the countries of Africa and put the African economic development agenda on hold.

J1 Cilliers, J.(2018). Africa must maximise the value of donor aid. Accessed December 20, 2018 through https://issafrica.org/iss-today/africa-must-maximise-the-value-of-donor-aid
J2 Karikari K (2002). Where Has Aid Taken Africa? Re-thinking Development. Accra, Ghana: Media Foundation for West Africa.
J3 Blanchard, O. & Leigh, D. (2013). Growth Forecast Errors and Fiscal Multipliers. IMF Working Paper
J4 Phillips, N. (2018) No Love Lost: A Brief History of Argentina and the IMF. Accessed December 22, 2018 through https://www.thebubble.com/history-argentina-imf/
J5 Ankrah, G. & Kwofei, R. (2015). Ghana to receive US$918M from IMF. Accessed December 22, 2018 throug http://www.ghana.gov.gh/index.php/news/1667-ghana-to-receive-us-918m-from-imf
J6 Buckley, R. P., & Fitzgerald, S. M. (2004). An Assessment of Malaysia's Response to the IMF during the Asian Economic Crisis. Sing J. Legal Stud., 96.
J7 Action Aid (2007). The IMF's Policy Support Instrument: Expanded Fiscal Space or Continued Belt-Tightening? Accessed December 22, 2018 through https://www.networkideas.org/doc/nov2007/ds28_psi_actionaid.htm

J8   CNBC (2013). Xi calls for China to 'stay the course': No one is in a position to dictate reform to US. Accessed December 28, 2018 through https://www.cnbc.com/2018/12/18/amid-trade-war-xi-jinping-says-china-must-stay-the-course-on-reform.html

CHAPTER TEN

The African Dream: The Promising Paths

Economic freedom cannot be sacrificed if political freedom is to be preserved.
—Herbert Hoover

High was the promise of the great African country that was the first to gain political freedom in Africa's liberation struggle. The country that vowed to the world that it was going to be an example that will prove to the world that the black man is capable of managing his affairs is now saddled with debt and sinking in poverty in the midst of her abundant resources because of mismanagement and unbridled greed. The "shit hole" comments and the hearsay calls for recolonization of the continent of Africa from outside the borders of Africa appear to show that the freedom fighters were wrong and that the black man is incapable of managing his affairs. The African dream now seems to be a nightmare.

There is not a better time for those who have not lost hope but still believe in the African dream to prove to the world that the founders of independent African countries were not wrong. They were not wrong when they founded the Pan-African Congress in Paris. They were not wrong when the women of Calabar and Owerri registered their protest. They were not wrong when they preferred to be jailed than to obey the orders of a foreign government. They were not wrong when they marched in protest against colonialism. They were not wrong when they said that the black man is capable of managing his affairs. They were not wrong when they shouted, "Freedom! Freedom!"

Many excellent suggestions have been made as to which is the best route for Africa to get back on track; the right path dreamed of by her oppressed citizens and shaped by her freedom fighters. Though laudable, the prescriptions offered are as many as the challenges facing the continent. This book has raised some issues for consideration. However, the suggestions might not be able to achieve the needed results or impact if certain fundamental issues are not corrected. Attaining the goal is not an easy task, but what is certain is that it will be more difficult in the coming years if actions are not taken to reverse current trends. Africa must get its politics right. Without this, it will be impossible to get things to work out. Politics determines everything. Politics is broader than governance itself, and the effect of politics on the whole system of governance is immeasurable.

IS THIS THE PATH?

I desire to see the day when an African country will be able to live up to the words of Kwame Nkrumah, "The black man is capable of managing his own affairs." Until then, I see the sweat and blood of the freedom fighters going to waste. Until then, I see the sacrifices of fellow Africans in the Diaspora, including Martin Luther King Jr. and Marcus Garvey, among others, turning to senseless sacrifices. But there are glimmers of hope. I can see a new Africa in the horizon. Faint but steady rays of lights are beginning to appear in the distant horizon. An Africa where its youth will no more ponder on how to cross the dangerous waters of the Mediterranean but creatively provide solutions to the world's problems and harness their country's resources to the envy of the world. I can only pray that these lights are not quenched over time just like the candles lit by the freedom fighters. These glimmers of hope are some of the few African countries that are beginning to show the way.

I can see Rwanda taking the paths that are leading to the pleasant places of sustained economic development. From a genocide some years ago,1994, that wiped out over eight hundred thousand of her population, the leadership is on progressive paths. Kigali, the nation's capital, is now the cleanest city in Africa, and if you choose to be corrupt, you better pray that you are not caught. I can see Ethiopia is squeezing through the grips of economic dependency to the honorable paths of economic development. In just six months after assuming office, a government led by a woman has drastically improved equality and freed political prisoners in pursuit of the ideals of democracy where people can politically agree to disagree. Her government is making it hard for corruption to thrive, and she is growing the economy in double figures.

My concern, my worries are the "voices" of dissent. These voices are familiar. Africa has heard these voices before and is beginning to hear them again. The voices that criticize the promising paths and raise resistance not from outside Africa but from among Africa's citizens. The misleading voices that make the crooked ways seem pleasant. Those voices are up again against the few governments that are showing the leadership and treading on the right paths. It is my prayer that Africa ignores the sweet melodic voices that call to destroy. I say this because these were the same voices that emerged against the likes of Kwame Nkrumah, Patrice Lumumba, and the others who, though had their weakness, were committed to the economic emancipation of Africa.

We must be appreciative of the fact that there cannot be a perfect government on this planet. Every government will have its flaws, so if a government is getting many things right, especially the basics, we must provide the support or constructive criticisms, but not stumbling blocks. This is the only way we can encourage good visions and initiatives to thrive on the continent of Africa. In our era, we hear voices that are saying that Pres. Paul Kagame is a dictator and the issues being raised are the issues regarding human rights. Yes, I don't support the abuse of human rights, but I equally do not support the actions of those who have purposed in their heart to oppose the economic development of a nation. Corruption in China is an offense punishable by death. To the Chinese, it might be the only punishment heavy enough to deter others from committing such crimes. The world has not been happy about China's stance on the issue of punishment meted out to corrupt officials, but I guess it is working out perfectly for the people of China.

Africa is in a fix. The donors that are pumping the taxes of their citizens into the African continent are discouraged because they are concerned that monies sent to Africa might end up in the pockets of corrupt politicians. Yet they raise issues when African leaders take harsh stands against corruption in the name of human rights. The frequent words of a Spanish American friend to me have always been on my mind: "You gotta do what you gotta do, man!" It is time for me to pass these words down to Africa as the continent struggles to disentangle herself from poverty. "You gotta do what you gotta do, Africa." It is time to do what must be done to pursue and attain prosperity for the land that is rich beneath its soil, and yet those whose feet tread on these riches have nothing to show to the rest of the world but poverty.

The good old book, the Bible, urges humankind to love their neighbors as ourselves. The countries or continents of the world can only love Africa as they love themselves, but they cannot love Africa more than the love they have for their countries. Africa needs to do whatever possible to get out of that "shit hole," and it is not going to be easy. Sacrifices have to be made. Endemic situations like corruption have to be rooted out by whatever means necessary. The leadership should have the will take to actions that might even cost them the next election if what they are doing is right. A popular Ghanaian proverb has it that "the actual length of the frog is known when it is dead." Some actions are understood or even considered destructive until the pieces come together to become complete. Kwame Nkrumah had his flaws just as every human being has, but his actions were not understood until he was overthrown by his people under the inspiration of external voices. A few decades later, he was remembered as voted as Africa's Man of the Millennium, and he is the only African leader whose statue is erected at the headquarters of the African Union in Addis Ababa.

"Enough, is enough!" so the saying goes. The shame that has been with Africa is enough. The ridicule is enough, and the abuse is enough. Africans need to rise and do what needs to be done. The leadership must feel the pressure and the heat from the people to do the right things, and no excuse or explanations need to be accepted.

A government is voted into power to make things better for their people, not to explain why they cannot perform. The governments that cannot deliver to the satisfaction of citizens must humbly step aside for those who can to form the government that will be responsive for the people. Government is the mandate from the people, not a privilege of the ruling class.

Everywhere Africans find themselves outside the African continent, they are treated with contempt. The leaders, because they fly in their planes or have diplomatic passports, should know that they will one day be stripped off that position and will have the same treatment. I am beginning to think that it is the reason they don't want to step down when they have served their term. I suggest the leaders of Africa, after they have served their terms, are handed with ordinary passports instead of the diplomatic passports so that they can feel what the average African feels and use their influence to continue to demand the right things to be done. I remember flying from Miami to Zurich through Philadelphia one time. As I went through immigration and baggage claim in Zurich, I was on my way out of the airport when I saw about five policemen chatting at a distance. I arrived at around 8:00 a.m., and the airport wasn't busy at the time. I kept moving, dragging my traveling bag along. As soon as one of the police officers spotted me—the only black man in the area at the time—he started coming straight toward me. I ignored him while heading toward the exit.

"My friend, where are you from?" he shouted as he was still some feet away.

"Philadelphia," I answered as I kept heading toward the exit.

At the mention of Philadelphia, he stopped and started moving back and remarked, "Enjoy your stay in Switzerland, sir!" he shouted.

"Thank you, sir!" I shouted back in response.

As I sat in the train from the airport to my final destination, I kept thinking over the drama that had just unfolded. After the scrutiny at the port of entry and being granted entry into a country, why should the African be picked among all the others for another series of questioning? I imagined what would have happened if I was on a flight from Accra, Ghana. I know for a fact I wouldn't have gotten the "enjoy your stay" wishes from the police officer. I would have had serious questions to answer. I have had such experiences over the years, and I do believe those who have traveled on ordinary African passports have had similar experiences. I was once in an arrival queue at Schiphol Airport in Amsterdam. When it got to my turn with the immigration officer in the booth, I answered so many questions for my ten days' visit—so much so that when I was finally cleared for entry, all the other lines including scores of people that were behind me were gone as they joined the other queues for entry. As I sat in the car that picked me up from the airport, the colleague who picked me up asked why I kept that long because he nearly went back home because he thought I was not on the plane. I told him that "getting a visa into your country and getting clearance into your country are two major tasks for the African. None of them are easy." If you are an African and not challenged by such treatments in and outside Africa, and you are not willing to see this change, then something must be seriously wrong somewhere.

Though the experiences out there in other countries for the African are concerning, I will never condemn any law enforcement officers of foreign governments for doing what they need to do to protect their citizens. I will rather condemn the governments of African countries for not doing what they need to do to protect their citizens and give them dignity in the eyes of the world. I can only believe that the African dream is a dream deferred, not a nightmare. A dream deferred by self-centered people whose desire is to be served and not to serve. A dream deferred with the support of the people who don't want to see Africa succeed. Africa has been dreaming for far too long, and it is time to wake up to the actualization of the dreams of her fathers. If this generation fails to find the economic formula that will heal the cancerous disease of poverty from the African continent, they should not be ashamed when future historians write of this generation as the people whose economic condition got worse while battling the disease of poverty with the pills of foreign aid while doling out its limited oxygen of natural resources.

In the United States, Canada, Switzerland, United Kingdom, and all over the world, you will find Africans among the dedicated working class. Many times, they are paid wages that they can hardly survive on. Yet they continue to do their best because back at home, they might be earning nothing because their economies are unable to create jobs for them. How long can Africans continue to be laborers and nation builders in other countries, while their countries lie in ruins? Africa, rise up! It will be desirable to see the fifteen-year-olds get to twenty-five and find work in African factories and tech companies instead of embarking on desperate journeys across dangerous seas and deserts.

WHERE TO BEGIN

As indicated earlier, African has to get her politics right as many share the opinion that the government is the party that must initiate the desired change. Governments need to lay a good foundation for a more prosperous and economically independent future. My doubts are with the way successive governments keep repeating the same mistakes and taking the same old paths that are leading our countries nowhere. If the government is failing in this direction, then the already weak civil society should step in to draw the attention of governments back to the basics and to educate the citizens on how to demand good services and better governance appropriately. And if the civil society fails, then the ordinary man has to make his voice heard.

Recently, commuters on the Accra to Tema motorway, realizing the potholes that have developed on the route, decided that they were no more going to pay the road tolls because they do not see the essence of continuing to pay for road tolls when the money that was meant to maintain the road has not been used for the purpose. It took that action on the side of the commuters to have the road fixed. While I do not call for people to disrespect law and order, I cite this as an example that if the citizens of a country will not stand up and will not speak up in some cases, they may not be able to get what they deserve. As the proverb in Akan goes, "If you don't speak up at the barbering salon, you are more likely to leave the salon with a bad haircut."

The youth should take the lead in bringing the change that they want to see for themselves. All over the world, it is evident that the youth, mainly from the ages of twenty to forty-five, are the ones that lead the fight for a social change. This is precisely so because the people in this age group do not only live to initiate the kind of change that they want but also live to see it happen in their lifetime. They are also the ones that will work because they have the energies to make the needed sacrifice to cause the desired changes to happen.

The people should join forces to oppose the systems that are suppressing liberties and impoverishing them further. They should rally together to change nonperforming governments that only appear to be working in election years and handing over peanuts in exchange for the power of the people. Maybe they have joined forces to change governments before, but they couldn't see any difference between the latter and the former. They see the present situation as the norm, and all they can do now is to take whatever little amount they can from the political class in exchange for the power to rule over them. But no matter how long an abnormal situation persists, it cannot become normal.

There is power in numbers, and it is about time the unemployed and the underemployed begin to exercise the power they have. The energies they put into searching for escape routes in the deserts to Europe and other parts of the world should be rechanneled into patriotism and activism for policy changes for a more progressive future. In the era of social media, it is easy to organize and make demands on the governments for the necessary actions to be taken to reverse a deteriorating situation. The African dream seems impossible, but so did Africa's independence some decades ago. Africa's economic independence is possible; what is needed are the people who have the commitment to break the barriers and push for it. The path to economic freedom is different from the paths of political freedom. Africa needs to switch from the old paths because they are paths to nowhere.

As the path to political independence is different from that of economic independence, so is the effort. The effort needed to actualize economic independence seems to be far higher than what was expended during the fight for political freedom. This may partly be because of the concessions made during the independence struggle. Many African countries made concessions that continue to work against the economic independence of their countries. In 1962, Sylvanus Olympio, the first President of Togo wanted to make changes in some of the independence concessions. Many believe that the move resulted in the overthrow of his regime in January 1963 by forces from fellow countrymen who were suspected of being mobilized and resourced by foreign governments. President

Olympio died in the hands of fellow countrymen while on a mission to actualize his country's political, economic and monetary freedom.

Many other governments have been toppled for similar offenses. In modern 'democratic' Africa, the trend continues as resources are mobilized at the blind side of the electorate to help vote 'ambitious' governments out of office. Could it be that many African leaders are not doing what they have to do to push forward the economic independence agenda because they don't want to offend foreign governments? If it is so, then the masses ought to know and maybe accept the fate that Africa's independence ended at her political independence. If the answer is negative, then African leaders ought to up their game by charting new and purposeful paths. I challenge African leaders and the vociferous Pan Africanists to move beyond the threats of electoral defeats or deaths and do what they ought to do for their countries. Because, whether they act or not, sacrifices are being made in one way or the other. Either the leaders or economic freedom fighters will be removed from office or the world in an attempt to do what is right for their people, or they will watch while poverty and diseases eliminate their countrymen from the earth. Africa has to choose the best of these sacrifices.

"Whatever form of government we adopt as a people to suit our peculiar circumstances, our basic tenet is our common yearning and concern for every individual; for politics, whatever its colour must be an avenue to serve our fellowmen."[K1] These are the words of former president Jerry John Rawlings, and he can never be far from right. If the people of Africa don't know where they are going, it will be easy for others to lure them on any path or accept any assistance offered. Countries and institutions are willing to assist Africa in diverse ways, but until Africa is convinced about where it is heading, it will never know which country in the world offers the true partnership or assistance that will take Africa to the destination of her dreams. Africa must chart and stick to new paths for the present paths are misleading; they are paths to nowhere.

K1 Obour, S. (2017). 10 powerful quotes by Jerry John Rawlings. Accessed December 15, 2018 through https://yen.com.gh/58293-powerful-quotes-jerry-rawlings.html#58293

REFERENCING

Chapter 1: Hope; the Inspirational path

A1
Abbey, R. A. (2018). The broken dream: 75% of Ghanaians want out of country – survey. Accessed November 22, 2018 through https://thebftonline.com/2018/business/the-broken-dream-75-of-ghanaians-want-out-of-country-survey/

A2
Nyabor, J. (2017). 1.7 million Ghanaians applied for US visa lottery in 2015. Accessed November 29, 2018 through http://citifmonline.com/2017/04/04/1-7-million-ghanaians-applied-for-us-visa-lottery-in-2015/

A3
Pew Research Center, (2018). At Least a Million Sub-Saharan Africans Moved to Europe Since 2010. Accessed November 22, 2018 through http://www.pewglobal.org/2018/03/22/at-least-a-million-sub-saharan-africans-moved-to-europe-since-2010/

A4
Happy Ghana (2018), Give me Ghana, Nigeria to recolonize; in a year they will become first-class countries - North Korean President. Accessed November 28, 2018 through https://www.happyghana.com/give-me-ghana-nigeria-to-recolonize-in-a-year-they-will-become-first-class-countries-north-korean-president/

A5
Heward Mills, D. (2011), Evangelism and Missions. Xulon Press.

A6
Myjoyonline, (2017), Full text: First independence speech by Kwame Nkrumah. Accessed December 1, 2018 through https://www.myjoyonline.com/news/2017/March-6th/full-text-first-independence-speech-by-kwame-nkrumah.php

Chapter 2: Resource waste

B1
Nkrumah, K. (1953) To-day we are here to claim this right to our independence, Motion of Destiny. Accessed December 1, 2018 through https://speakola.com/political/kwame-nkrumah-motion-of-destiny-independence-1953

B2
Ghanaweb, (2016). The modern conspiracy against Ghanaians. Accessed December 1, 2018 through https://www.ghanaweb.com/GhanaHomePage/features/The-modern-conspiracy-against-Ghanaians-494097

B3
Kwawukume, S. (2018). Revised Seven Years Of Oil Production In Ghana -An Independent Report. Accessed December 1, 2018 through https://www.modernghana.com/news/879427/revised-seven-years-of-oil-production-in-ghana-an-independe.html

B4
Eigen, P. (2018, December 31). DW Journal Interview

B5
Smith, E. E. (1991). From concessions to service contracts. *Tulsa LJ*, *27*, 493.

B6
United Nations, (2002). Security Council is told peace in Democratic Republic of Congo needs solution of economic issues that contributed to conflict. Accessed December 1, 2018 through https://www.un.org/press/en/2002/sc7547.doc.htm

B7
United Nations (2004). United Nations Commission Mission to the Democratic Republic of Congo, Report on the conclusions of the special investigation concerning allegations of summary executions and other human rights violation perpetrated by the Armed Forces of the Democratic Republic of Congo (FARDC) in Kilwa (Katanga Province) on 15th October 2004). Accessed December 28, 2018 through https://www.ohchr.org/Documents/Countries/CD/LikofiReportOctober2014_en.pdf

B8
Deaton, A., & Miller, R. I. (1995). *International commodity prices, macroeconomic performance, and politics in Sub-Saharan Africa.* Princeton, NJ: International Finance Section, Department of Economics, Princeton University.

B9
Deaton, A. (1999). Commodity prices and growth in Africa. *Journal of Economic Perspectives, 13*(3), 23-40.

B10
Sarkar, S. & Pema, T. (2017). Tullow says makes oil discovery in Kenyan well. Reuters. Accessed December 10, 2018 through https://www.reuters.com/article/us-tullow-exploration/tullow-says-makes-oil-discovery-in-kenyan-well-idUSKCN18D0SM

B11
Ng'wanakilala, F. (2017). Tanzania makes big onshore natural gas discovery - local newspapers. Reuters. Accessed December 10, 2018 through https://www.reuters.com/article/tanzania-gas/tanzania-makes-big-onshore-natural-gas-discovery-local-newspapers-idUSL8N16427G

B12
Vanguard, (2015).Nigeria discovers 44 mineral deposits in 500 locations. Accessed December 10, 2018 through https://www.vanguardngr.com/2015/11/nigeria-discovers-44-mineral-deposits-in-500-locations/

B13
Musisi, F. (2013). New minerals worth trillions discovered in Busoga region. https://www.monitor.co.ug/Business/Commodities/New-minerals-worth-trillions-discovered-in-Busoga-region/688610-2021266-by9acv/index.html

B14
Diallo, T. & Jabkhiro, (2018). Mali to produce lithium by 2020 with 694,000 T discovered.
Accessed December 10, 2018 through https://www.reuters.com/article/mali-lithium/mali-to-produce-lithium-by-2020-with-694000-t-discovered-idUSL8N1XU2MF

B15.
ANGOP (2017), Rwanda: New Minerals Found as Government Steps Up Exploration. Accessed November 30, 2018 through http://www.angop.ao/angola/en_us/noticias/africa/2017/1/7/Rwanda-New-Minerals-Found-Government-Steps-Exploration,3324b1b7-1391-4670-abe6-e8fa2dcf8de3.html

B16
Stiglitz, J. E. (2012). From Resource Curse to Blessing. Project Syndicate. Accessed November 30, 2018 through https://www.project-syndicate.org/commentary/from-resource-curse-to-blessing-by-joseph-e--stiglitz?barrier=accesspaylog

CHAPTER 3 - Governance Institutions

C1
Baker, P. (2009). Obama Delivers Call for Change to a Rapt Africa. New York Times. Accessed December 15, 2018 through https://www.nytimes.com/2009/07/12/world/africa/12prexy.html

C2
Hanson, S. (2009). Corruption in Sub-Saharan Africa. Accessed December 15, 2018 through https://www.cfr.org/backgrounder/corruption-sub-saharan-africa

C3
McGowan, P. J. (2003). African military coups d'état, 1956-2001: Frequency, trends and distribution. The Journal of Modern African Studies. 41. 339 - 370. 10.1017/S0022278X0300435X.

C4
Rickard. C. (2018). Lesotho: Independence of the Judiciary in Peril. Accessed December 15, 2018 through https://africanlii.org/article/20180518/lesotho-independence-judiciary-peril

C5
Transparency International (2007). Global Corruption Report 2007: Corruption in Judicial Systems. Cambridge University Press.

C6
Themnér, A. (Ed.). (2017). *Warlord democrats in Africa: ex-military leaders and electoral politics.* Zed Books Ltd..

C7
Banjo, A. (2008). THE POLITICS OF SUCCESSION CRISIS IN WEST AFRICA: THE CASE OF TOGO. *International Journal on World Peace, 25*(2), 33-55. Retrieved from http://www.jstor.org/stable/20752832

C8
Wikipedia (2016). 2016 Equatorial Guinean presidential election. Accessed December 15, 2018 through https://en.wikipedia.org/wiki/2016_Equatorial_Guinean_presidential_election

C9
Orji, N. (2009). CIVIL SOCIETY, DEMOCRACY AND GOOD GOVERNANCE IN AFRICA. CEU Political Science Journal, 4(1).

C10
Gyimah-Boadi, E. (1996). Civil society in Africa. *Journal of Democracy, 7*(2), 118-132.

C11
Chabal, P., & Daloz, J. P. (1999). Africa works: Disorder as political instrument (African Issues). *James Currey, Oxford.*

CHAPTER 4 - Lawlessness

D1
Edmunds, S. E. (1924). The lawless law of nations. . *Louis L. Rev., 10*, 171.

D2
Šimonović, I. (2017). Lawlessness in the Heart of Africa. Accessed December 15, 2018 through https://www.huffingtonpost.com/ivan-simonovic/lawlessness-in-the-heart-_b_3755302.html

D3
Brainy Quotes, (2017). John F. Kennedy Quotes. Accessed December 15, 2018 through https://www.brainyquote.com/quotes/john_f_kennedy_125480

D4
Allen, J. (2012). *Above Life's Turmoil: eBook Edition*. Jazzybee Verlag.

D5
GWS Online (2018). President John Evans Atta-Mills storms Tema Port. Accessed December 15, 2018 through https://www.ghanawebsolutions.com/videos.php?v=ed57WhR5ouQ

CHAPTER 5 - Dependency Mindset

E1
Connors, W. (2010). China Extends Africa Push With Loans, Deal in Ghana. Accessed December 12, 2018 through

https://www.wsj.com/articles/SB10001424052748703384204575509630629800258

E2

Nyavor, G. (2018). Ghana considering floating rare $50bn Century Bond. Myjoyonline.com. Accessed December 15, 2018 through https://www.myjoyonline.com/business/2018/september-3rd/ghana-considering-floating-rare-50bn-century-bond.php

E3

Ghanaweb (2018). Cash for Bauxite: Parliament okays $2bn Ghana-China barter trade. Accessed December 12, 2018 through https://www.ghanaweb.com/GhanaHomePage/NewsArchive/Cash-for-Bauxite-Parliament-okays-2bn-Ghana-China-barter-trade-673389

E4

Aljezeera (2018). China's Xi offers $60bn in financial support to Africa. Accessed December 12, 2018 through https://www.aljazeera.com/news/2018/09/china-xi-offers-60bn-financial-support-africa-180903100000809.html

E5

AllAfrica (2018). Africa: South African President Rejects Claims of Chinese "Colonialism"
Accessed December 15, 2018 through https://allafrica.com/stories/201809040284.html

E6

Later, V., & Mususa, P. (2018). Is China really to blame for Zambia's debt problems? Accessed December 11, 2018 through https://www.aljazeera.com/indepth/opinion/china-blame-zambia-debt-problems-181009140625090.html

E7

Ebatamehi, S. (2018). Zambia Deported Renowned Pan-African, Prof. PLO Lumumba, for Speaking Up Against China. Accessed December 12, 2018 through https://www.africanexponent.com/

post/9177-kenyas-professor-patrick-lumumba-denied-entry-into-zambia-deported-back-to-kenya

E8
AZ Lyrics (2018). Lucky Dube Lyrics. Accessed December 16, 2018 through https://www.azlyrics.com/lyrics/luckydube/victims.html

CHAPTER 6 - Education

F1
Berman, E. H. (1971). American influence on African education: the role of the Phelps-Stokes Fund's Education Commissions. *Comparative Education Review*, 15(2), 132-145. http://www.jstor.org/stable/1186725.

F2
Forster, P. (1965). Education and Social Change in Ghana. Routledge & Kegan Paul. London

F3
Prices Ghana (2019). History Of Education in Ghana. Accessed February 7, 2019 through
https://pricesghana.com/history-of-education-in-ghana/

F4
Republic of Ghana (1995). Ghana Vision 2020 (The First Step 1996-2000). Presidential
Report on Coordinated Programme of Economic and Social Development Policies
(Policies for the Preparation of 1996-2000 Development Plan).

F5
IMF (2012). South Africa: 2012 Article IV Consultation. IMF Country Report No. 12/247. Accessed January 2, 2019 though https://www.imf.org/external/pubs/ft/scr/2012/cr12247.pdf

F6
Ghanaweb (2018). Wage bill hits GHC14bn, takes 45% of tax revenue in 2017. Accessed December 11, 2018 through https://www.ghanaweb.com/GhanaHomePage/business/Wage-bill-hits-GHC14bn-takes-45-of-tax-revenue-in-2017-671387

F7
Clements, B., Gupta, S., Karpowicz, I. & Tareq, S. (2010) Evaluating Government Employment and
Compensation. Washington DC, International Monetary Fund, Fiscal Affairs Department.

F8
Herbst, J. (1989). The creation and matintenance of national boundaries in Africa. International Organization, 43(4), 673-692. doi:10.1017/S0020818300034482

F9
BBC (2011). Rwanda: How the genocide happened. Accessed December 11, 2018 through https://www.bbc.com/news/world-africa-13431486

F10
European Commission (2017). About multilingualism policy. Accessed December 16, 2018 through https://ec.europa.eu/education/policies/multilingualism/about-multilingualism-policy_en

CHAPTER 7- CORRUPTION

G1
Wikipedia (2019). Capital punishment by country. Accessed December 16, 2018 through https://en.wikipedia.org/wiki/Capital_punishment_by_country

G2
Acton, H. (1961). "Lord Acton". Chicago Review. 15 (1): 31–44. doi:10.2307/25293642.

G3
Denny, C. (2004). Suharto, Marcos and Mobutu head corruption table with $50bn scams. Accessed December 16, 2018 through https://www.theguardian.com/world/2004/mar/26/indonesia.philippines

G4
Richburg, K. B. (1991). Mobutu: A rich man in poor standing. Accessed December 16, 2018 through https://www.washingtonpost.com/archive/politics/1991/10/03/mobutu-a-rich-man-in-poor-standing/49e66628-3149-47b8-827f-159dff8ac1cd/?noredirect=on&utm_term=.c05200a56440

G5
Smith, S. (2015). President Mobutu's ruined jungle paradise, Gbadolite - in pictures. Accessed December 16, 2018 through https://www.theguardian.com/cities/gallery/2015/feb/10/inside-gbadolite-mobutu-ruined-jungle-city-in-pictures

G6
Daily Graphic, (1961). Bribery - Our Main Problem, says Busia. Monday, March 22, 1971

CHAPTER 8 - Technology and Data Management

H1
Simons, G. (2014). The Airbus A380: A History. Pen & Sword Books

H2
Drucker, P. (2012). The practice of management. Routledge.

H3
Ghanaweb (2018) Ghana to spend 1.22 billion dollars on national ID cards – NIA reveals. Accessed December 16, 2018 through

https://www.ghanaweb.com/GhanaHomePage/NewsArchive/Ghana-to-spend-1-22-billion-dollars-on-national-ID-cards-NIA-reveals-654375

H4
The Finder (2018). Rising recklessness on roads 1,212 Killed in 6 months. Accessed December 18, 2018 through https://www.thefinderonline.com/news/item/13549-rising-recklessness-on-roads-1-212-killed-in-6-months

H5
Quoteswise (2018). 411 Martin Luther King Jr Quotes. Accessed December 18, 2018 through
http://www.quoteswise.com/martin-luther-king-jr-quotes-8.html

H6
Galetto, M. (2016). What is Data Management? Accessed December 18, 2018 through
https://www.ngdata.com/what-is-data-management/

H7
Ombati, C. (2013). Uhuru unveils new traffic command centre. Accessed December 20, 2018 through
https://www.standardmedia.co.ke/article/2000099808/uhuru-unveils-new-traffic-command-centre

CHAPTER - 9 Donor Partners

J1
Cilliers, J.(2018). Africa must maximise the value of donor aid. Accessed December 20, 2018 through https://issafrica.org/iss-today/africa-must-maximise-the-value-of-donor-aid

J2
Karikari K (2002). Where Has Aid Taken Africa? Re-thinking Development. Accra, Ghana: Media Foundation for West Africa.

J3
Blanchard, O. & Leigh, D. (2013). Growth Forecast Errors and Fiscal Multipliers. IMF Working Paper

J4
Phillips, N. (2018) No Love Lost: A Brief History of Argentina and the IMF. Accessed December 22, 2018 through https://www.thebubble.com/history-argentina-imf/

J5
Ankrah, G. & Kwofei, R. (2015). Ghana to receive US$918M from IMF. Accessed December 22, 2018 throug http://www.ghana.gov.gh/index.php/news/1667-ghana-to-receive-us-918m-from-imf

J6
Buckley, R. P., & Fitzgerald, S. M. (2004). An Assessment of Malaysia's Response to the IMF during the Asian Economic Crisis. Sing J. Legal Stud., 96.

J7
Action Aid (2007). The IMF's Policy Support Instrument:Expanded Fiscal Space or Continued Belt-Tightening? Accessed December 22, 2018 through https://www.networkideas.org/doc/nov2007/ds28_psi_actionaid.htm

J8
CNBC (2013). Xi calls for China to 'stay the course': No one is in a position to dictate reform to US. Accessed December 28, 2018 through https://www.cnbc.com/2018/12/18/amid-trade-war-xi-jinping-says-china-must-stay-the-course-on-reform.html

CHAPTER 10 - The African Dream

K1
Obour, S. (2017). 10 powerful quotes by Jerry John Rawlings. Accessed December 15, 2018 through
https://yen.com.gh/58293-powerful-quotes-jerry-rawlings.html#58293